# Tourism, Philanthropy and School Tours in Zimbabwe

This book explores the phenomena of school tours and tourism. It explores tensions of authenticity and artificiality in the school site, being both a place of community learning and a spectacle for tourism consumption.

Through the example of a school in Matabeleland North, Zimbabwe, the book examines the act of a school tour, whose main aim is to providing fund for the school. It offers a unique interdisciplinary lens that examines both the school as a tourism destination and as a site of learning. By drawing on these two fields, the book provides insights into the tensions inherent in a school that is also a tourism destination. This book will demonstrate to readers the tensions present in tourism partnerships with schools that include some source of philanthropic funding and unpack the complexities of tourism that draws on stereotypical cultural images. It explores these tensions through the lens of school leaders, students, teachers, and tourism personnel.

The book provides a major and unique contribution to the field of tourism studies and education. It will be of interest to students and researchers interested in tourism studies, sociology, education, philanthropy, development studies, and the Global South.

**Kathleen Smithers** is a Lecturer in the School of Education at Charles Sturt University, Australia. Her main research interests are in sociology of education, philanthropy in education, and precarity in higher education.

## Routledge Insights in Tourism Series
Series Editor: Anukrati Sharma
*Head & Associate Professor of the Department of Commerce and Management at the University of Kota, India*

This series provides a forum for cutting edge insights into the latest developments in tourism research. It offers high quality monographs and edited collections that develop tourism analysis at both theoretical and empirical levels.

**Overtourism and Cruise Tourism in Emerging Destinations on the Arabian Peninsula**
*Manuela Gutberlet*

**Pseudo-Authenticity and Tourism**
Preservation, Miniaturization, and Replication
*Jesse Owen Hearns-Branaman and Lihua Chen*

**Developing Industrial and Mining Heritage Sites**
Lavrion Technology and Cultural Park, Greece
*Taşkın Deniz Yıldız*

**Innovation Strategies and Organizational Culture in Tourism**
Concepts and Case Studies on Knowledge Sharing
*Edited by Marco Valeri*

**Tourism and Poverty Alleviation in Nature Conservation Areas**
A Comparative Study Between Japan and Vietnam
*Nguyen Van Hoang*

**Sport Tourism, Events and Sustainable Development Goals**
An Emerging Foundation
*Edited by Anukrati Sharma, Miha Lesjak and Dusan Borovcanin*

**Tourism, Philanthropy and School Tours in Zimbabwe**
Problematising "Win-Win" Discourses
*Kathleen Smithers*

For more information about this series, please visit: www.routledge.com/Routledge-Insights-in-Tourism-Series/book-series/RITS

# Tourism, Philanthropy and School Tours in Zimbabwe

## Problematising "Win-Win" Discourses

**Kathleen Smithers**

Routledge
Taylor & Francis Group

LONDON AND NEW YORK

First published 2025
by Routledge
4 Park Square, Milton Park, Abingdon, Oxon OX14 4RN

and by Routledge
605 Third Avenue, New York, NY 10158

*Routledge is an imprint of the Taylor & Francis Group, an informa business*

© 2025 Kathleen Smithers

*British Library Cataloguing-in-Publication Data*
A catalogue record for this book is available from the British Library

*Library of Congress Cataloging-in-Publication Data*
Names: Smithers, Kathleen, author.
Title: Tourism, philanthropy and school tours in Zimbabwe : problematising "win-win" discourses / Kathleen Smithers.
Description: Abingdon, Oxon ; New York, NY : Routledge, 2025. | Series: Routledge insights in tourism | Includes bibliographical references and index.
Identifiers: LCCN 2024015497 (print) | LCCN 2024015498 (ebook) | ISBN 9781032527079 (hardback) | ISBN 9781032527086 (paperback) | ISBN 9781003407980 (ebook)
Subjects: LCSH: Tourism--Zimbabwe. | School sites--Zimbabwe. | Education--Zimbabwe--Finance. | Heritage tourism--Zimbabwe.
Classification: LCC G155.Z55 S65 2025 (print) | LCC G155.Z55 (ebook) | DDC 306.4/819096891--dc23/eng20240510
LC record available at https://lccn.loc.gov/2024015497
LC ebook record available at https://lccn.loc.gov/2024015498

ISBN: 978-1-032-52707-9 (hbk)
ISBN: 978-1-032-52708-6 (pbk)
ISBN: 978-1-003-40798-0 (ebk)

DOI: 10.4324/9781003407980

Typeset in Times New Roman
by SPi Technologies India Pvt Ltd (Straive)

# Contents

# Acknowledgement

I would like to thank the Matopo School (a pseudonym) community for welcoming me and supporting me during my time in Zimbabwe. The participants in this research shared their lives and their interests with me, and I will always be thankful for their generosity. The research in this book was also supported by an Australian Government Research Training Program (RTP) Scholarship and Fee Offset.

# 1 Introduction

Imagine, for a moment, the set of a movie filmed in an unnamed country in Africa. As the opening scene begins, the country melds into the homogenous continent 'Africa' and seems alluringly exotic. As you expect, red dirt and dry, sparse, savannahs stretch as far as the eye can see. Dotted across the landscape are animals: elephants, lions, giraffes, and zebras. There are no tall buildings, no sign of modern technology. Having established the scene, the camera pans across as the main character, a white American, has come to visit a rural school. In the foreground, the man stands in a red dirt playground and his skin is slightly pink in the sun. Dressed in khaki, he surveys the landscape with a camera at the ready. In his hands, there are packages of pens and pencils which are gleefully given to smiling children. After the gift exchange, the children spontaneously begin to dance, which the main character enthusiastically photographs, with plans to upload the smiles and laughter to his social media timeline as soon as the group returns to their pristine, expensive hotel. He will return with stories about how the children were so happy, despite having so little material possessions.

The above description could represent one of many movies, books, or tourism advertising which depict the 'real Africa' (I have discussed this elsewhere; see Smithers, 2023). This image of Africa is not the image I was raised with. I was born in Zimbabwe and lived there for a few short years as a young child. From this time, my parents had video footage of Harare, the capital city, and pictures of our house which was distinctly *not* a thatched hut. I was raised on stories of the time my parents spent in Zimbabwe but also on stories of turmoil, economic problems, and hardship. As a child growing up in Australia, it was a fun fact to tell other children that I was from Zimbabwe, a country in Africa! I remember when the quintessential 2000s movie *Mean Girls* entered popular culture – it offered a very stereotypical view of what it meant to live in Africa (no country is specified) and yet, felt so familiar as one character says to the other, 'so, if you are from Africa, why are you white?'.

As a student at university, I visited Zimbabwe on a study tour. Notably, the coordinator of this study tour was highly sensitive to the need to present multiple images of Africa, to not be a 'white saviour' and prepared us to visit Zimbabwean schools to learn, rather than to teach or take over. The focus was on our learning, rather than any white saviour images of 'helping', although it

DOI: 10.4324/9781003407980-1

could be argued that the use of a school in another country to learn is problematic in itself. After the study tour, I returned to Australia with my own image of Zimbabwe both enriched and complicated – I was now confronted with a story of disparate circumstances, of extreme wealth and extreme poverty. At a family gathering not long after I returned, somebody said to me, 'Wow, your pictures looked great of your trip. So great that you were there to help those teachers. Were you teaching them to teach?'. I was deflated, despite trying to avoid paternalistic white saviour behaviours, this was still the dominant expectation those around me had. I realised there were dominant images that we cannot escape.

The study tour was formative for me, not only as it began to help me see that my image of Africa was different to that of most of my peers, but also as it provided the catalyst for the research project that forms the basis of this book. While on the study tour, I was observing a lesson by a teacher in a primary school in Zimbabwe when suddenly the class were on their feet greeting a white woman and man who had entered the room, and these visitors had a pair of shoes that they gave to a child. It was a short disturbance to the class and I later asked the teacher what had happened. These tourists had been part of a larger tour group that had visited the school for a few hours. The tourists had no prior association with the student who was given the shoes, rather they had looked for a student they felt 'needed it'. My class had not been disrupted by the group except for the short visit by the couple, but the Grade 7 class were required to greet and entertain the tourists. At the time, I was horrified. Somewhat naively, I did not also see myself as a temporary interloper in the school.

Since this encounter, I have puzzled over how this type of tourism comes to be in a school, and under what conditions teachers will accept this type of tourism. The type of tourism I witnessed on the study tour and the type of tourism that is detailed in this book is what I would call a 'school tour', in which tourists visit a school for a short period of time (2–3 hours). They may have a prior association with the school, but also may not. They participate in cultural displays as part of the tour, hear the school history, and visit classrooms. In considering the type of tourism involved in a school tour, there are several different types of tourisms that seem to intersect and cover school tours and visits to schools. As someone experienced in the field of education and not tourism, journeying through these multiple tourisms that overlay and intersect enabled further insight into the field of tourism studies (see Chapter 4 for more detail).

It appears at the crux of it, many activities that involve schools are an exercise of corporate social responsibility. Corporate Social Responsibility (CSR) is touted as an answer to the tourism industry's need to 'give back' to communities, particularly with more recent calls for the tourism industry to alleviate damage to local communities. CSR involves a range of activities such as donations of goods and services, environmental programmes, and financial support of workers. There are many types of businesses who donate to schools as part of CSR activities. When schools are involved, the relationship and definition of this tourism can become blurred. Alongside CSR, there are several types of tourism that include visits to schools such as developmentourism, volunteer tourism, village tourism, philanthropic tourism, philanthrotourism, slum tourism, and more (I provide further detail in Chapter 4).

Given the current demarcation of tourisms into smaller and smaller niches, it is necessary to name the tourism encountered in this book. I have chosen the term developmentourism as the most fitting, although imperfect term. Baptista (2011) coined the term 'developmentourism' to reflect the 'integration of "development" discourse, knowledge and action into the tourism experience' (p. 663). Although Baptista examined the work of developmentourism in a broader context, it is a useful term for this book as it provides an umbrella term in which the school tour can sit (see also Smithers & Ailwood, 2022). As I will detail in later chapters, the tourism at the school discussed in this book (Matopo School – a pseudonym) is based in a need to develop and further the local area. Although it has a development and educational focus, it began and continues to be both a development initiative and a distinct tourism product.

To return to the story that opened this book, I offer my own experience not to centre myself in the narrative, but to provide insight into my own interest and connection to Zimbabwe. In the rest of the book, I use the metaphor of a movie set to frame the analysis. I use this metaphor and analysis to unpack the varying ways the image of Africa is presented, represented, and used by one school in Zimbabwe: Matopo School. In doing so, I speak back to the common refrain within tourism studies, that of a simple relationship between host and guest that lacks agency and power.

## Image(s) of Africa

As I highlighted above, I have always been particularly struck by the concept of the 'image of Africa' which highlights the held stereotypes, images, and tropes that govern Western ideas of Africa. In examining the idea of 'Africa', I utilise the concept of heterotopia to frame the discussion. Foucault (1967/1984) described heterotopia as, 'simultaneously mythic and real contestation of the space in which we live' (p. 24). Extending on this description, a heterotopia exists outside temporality: 'it is a real place but also one without place in the regular order of places, a place without place. It exists in real time but out of regular time, a time outside time' (Masschelein, 2018, p. 194).

As with many conceptualisations of space, it is hard to pinpoint exactly what heterotopia is. Heterotopias are unusual places which contest a society, whilst also being a potential sum of it. They can be both a real place and an idea of a place. One succinct explanation describes heterotopia as, 'heterotopia is an embodied and, in fact, thematized version of what society imagines and necessitates' (Hong & Vicdan, 2016, p. 122). I use heterotopia to understand the ways in which there is a 'thematized' version of Africa in tourism. The African continent is a heterotopia in Minority World imaginaries – any acts of culture within tourism ventures are constructed as a representation of all real sites within African cultures and there are contradictory elements which serve to represent Africa as an exotic 'dreamland'.

It is through the dissemination and exotification of an 'image of Africa' that tourism promotes and furthers neo-colonial interest in southern Africa. The term, 'image of Africa' was coined by Mudimbe (1988), who identified that there is a mythical place of 'Africa' situated in the hearts and minds of those from the Minority World. The image of Africa works to govern as a discourse and as a spatial rendering of heterotopia – specifically time stands still in images of Africa. When village life is depicted, there are thatched huts, half-naked or traditionally dressed women, and red dirt as the main features (Crossley, 2012). In the scenes which depict the pristine landscape, there are savannahs filled with animals, usually the big five: rhino, buffalo, lion, leopard, and elephant (Garland, 2008; Hays, 2012). In images of hospitality, white linen tents and tablecloths stand out against the brutality of the landscape with local people greeting the tourists with a smile (Garland, 2012). People within this image are always black, or they appear as the mythic white hunter who takes control of the landscape (Brandt & Josefsson, 2017). This image creates an illusion that defies reality, in which the cities within the continent are excluded (Baptista, 2011). In this image, time stands still. Further, the image is at once a flattened caricature of the continent, but it also shifts as needed. For example, in some depictions there is an element of danger to Africa and at the same time a level of poverty which requires Minority World intervention. Children are depicted as innocent and needing protection (Sheridan et al., 2019), but also as child soldiers (Martins, 2011).

An image of Africa is perpetuated through advertising (McClintock, 1995; Rideout, 2011) and media (Staples, 2006). These images of Africa ignore at least '50 years of independencies, urbanisation, industrialisation and the emergence of some of the most energetic metropolises in the world' (Baptista, 2011, p. 658). The image of a traditional way of life as simplistic, happy, and non-modern is a Minority World discourse intended to mask the extreme poverty that has resulted from the complex neo-colonial economic interventions from institutions such as the World Bank. Images of Africa are linked to the colonial past in which a romantic discourse of the continent was privileged (Landau, 2002). McClintock (1995) describes this romanticisation as,

> Neo-colonial nostalgia for an era when European women in brisk white shirts and safari green supposedly found freedom in the empire: running coffee plantations, killing lions and zipping about the colonial skies in aeroplanes – a misbegotten commercialisation of white women's 'liberation'.
>
> (p. 15)

As McClintock identifies, there is a certain type of uniform associated with the image of Africa: brisk, white shirts and safari green. These images of Africa govern the white identity as much as they govern the image of the Other – both the tourist and the host communities are discursively positioned in certain ways when they encounter tourism destinations.

Another element of this image is the idea that one demonstration of culture can represent all other African cultures. For example, the Maasai are commonly associated with representations of Africa (Hays, 2012). The Maasai are frozen in time, in an imagined wilderness, consumed by the Minority World through advertising and other media. The Maasai have a particular image associated with them, tall men in red blankets who stand surrounded by cattle and mud huts, which means that some in the Minority World can recognise the Maasai by name (Gardner, 2016; Hays, 2012). Indeed, the Maasai are 'revered in the West for their independence and perceived resistance to Western norms and values' (Gardner, 2016, p. 33). This discursive construction of a timeless and homogenised 'African' culture disguises the ongoing effects of colonial and neo-colonial interventions into the area in which the Maasai live (Gardner, 2016). Tourism furthers the idea that an experience with one African culture can represent *all* African cultures.

Baptista (2011) argues that the image of Africa is a discursive tool that both forms and shapes the reality in which people live. Images of Africa govern by distorting and shaping the social reality of what is Africa, and who can be African. hooks (1992) identifies the fantasy as a colonial longing that finds itself in a 'location be a country or a body, a dark continent or dark flesh, perceived as the perfect embodiment of that possibility' (p. 27). To relate this back to the heterotopia, the Minority World sustains a romantic fantasy of the primitive and continuously seeks a paradise that is linked to the imperial romantic fantasy; however, the fantasy exists only as a heterotopia. Heterotopias are an imaginative space and Foucault linked them to the imagination of children, describing them as 'realised utopias' which Johnson (2013, p. 798) describes as tangible heterotopias.[1]

## Education challenges

To foreground discussions of funding arrangements that follow, it is important to understand how the school in this research sits within the Zimbabwean school system. The schooling system suffered greatly during the years following Independence, due to economic and political instability alongside the extreme increase in student enrolments.

During colonial administration it was mostly white middle-class children who proceeded through to secondary schooling due to costly school fees and admission requirements that prevented non-white students from progressing (Abraham, 2003). After independence in 1980, the government begun policies with the intention of education becoming open and free for all Zimbabweans. In a method of reversing the effects of colonial policies, children older than primary school age were able to enrol in primary school. As a result, there was a large increase in the number of students, some older than primary school age into the system. Prior to independence, the system was designed for roughly 60% of the primary school age population. The increase meant a troubling transition for the system.

Alongside the increase in student numbers, the previous system had relied on private, selective schools as a means for providing secondary education. Whilst the existing primary school structure struggled to adjust to the rise in student enrolments; the secondary system had an unsuccessful transition, as there was simply not enough textbooks, classrooms, or teachers available (Chitiyo et al., 2010). The system struggled to rapidly develop the required infrastructure, instructional materials and relied upon teachers colleges to produce teachers quickly (Madziyire, 2015; Riddell, 1998).

Alongside the economic collapse, and period of hyperinflation in Zimbabwe, there are persistent crises for the education system today. Presently, many schools encounter deficiencies of furniture, other foundational infrastructure, and curriculum resources, particularly in rural areas (Mangwaya et al., 2013). As a solution to the lack of classroom buildings and desks, schools use 'hot-sitting', in which half the students participate in lessons in the morning and the other half in the afternoon; or one site may serve both as primary and secondary school (Dzimiri & Runhare, 2012). To make up for the lack of government funding and infrastructure, schools have begun to charge fees to cover teacher salaries or for the purchase of textbooks. However, students cannot technically be turned away for not paying these fees.

While funding for these resources should be provided by the government in government schools, they are often lacking or simply not there – with a large proportion of the budget for education allocated to teacher wages (Hlupo & Tsikira, 2012). Regardless of the budgetary provisions, government teacher salaries, if they are paid, are just on the poverty line at roughly US$6000 a year (Utete-Masango, 2016).

Satellite schools were designed as an interim solution to help alleviate the additional schools needed in land reform areas. They have since become an enduring characteristic of the Zimbabwe school system (Mudzingwa & Magudu, 2013; Tarisayi, 2019). Satellite schools are usually built and funded by the community, with the government sometimes providing teachers to the community (Tarisayi & Manik, 2019). Although not government-funded schools, satellite schools must follow Minsitry of Primary and Secondary Education's (MOPSE) curriculum and policies. Matopo School is one such school. Two of the ten teachers were employed by MOPSE and the school was in the process of gaining official registration with the education department.

Satellite schools' buildings and other vital resources are not funded by the government (Mangena & Ndlovu, 2013), some aim to obtain contributions or assistance from donors and NGOs (Magudu, 2020; Sithole, 2017; Tarisayi, 2019). Going hand-in-hand with donations, some schools ask foreign supporters to fund individual students through sponsorship schemes (Dekeza, 2018). The payment of school fees is an important factor that prevents students from dropping out of school (Madziyire, 2015). As I will discuss in later chapters, with the rise of tourism-oriented NGOs in Zimbabwe, and other charitable ventures, schools have acquired additional possibilities for funding, and they are becoming more inventive in the method in which they promote themselves.

There is no doubt that schools are showing ingenuity in their response to the continuing under-resourcing of schools and economic crises. The type of tourism and the creation of Matopo School itself are examples of the Zimbabwean term *kuronga-ronga* (make a plan), in which schools and communities are creating grassroots approaches to support the education system (Jones, 2010).

## A note on terminology

I use the term 'race trouble' to explore concepts of whiteness and race relations in this book. Discussions of race and racism can be fraught in all societies, but further so in countries historically dominated by white ways of thinking and doing. An aim of analysing 'race trouble' is to study 'how people account for their use of common-sense representations by describing their actions and their situational contexts' (Durrheim et al., 2011, p. 107). In their theorisation of race trouble, Durrheim et al. (2011) argue for a recognition that moving beyond binaries of racist/non-racist can be fruitful to form an analysis of how discourses and power relations work to subjugate.

Despite some prevailing post-racial colour blindness claims, we 'do race' in every facet of our lives, from walking on the footpath to commuting to work (Durrheim et al., 2011; Haslanger, 2019). The 'doing' of race includes the embodiment of racialised practices in the way we stand, move, and interact with each other (Ahmed, 2007; Fanon, 1986). Racialised discourses should be considered within an analysis of *practices* by asking, 'how such discursive representations fit into and form part of the practical contexts of race trouble?' (Durrheim et al., 2011, p. 132). Theorising of 'race trouble' is, therefore, based around the interrogation of practices that are not explicitly about race but an interrogation of practices that are (co)incidental formations which reproduce racism (Durrheim et al., 2011). I use this as a basis of understanding race as beyond the overt racist practices of those usually marked as 'racist'. Rather, I interrogate the racialised practices which characterise 'race trouble' and the dilemmas of interpretation that surround these ideas.

In this book I make use of the terms 'Majority World' and 'Minority World' in reference to the countries sometimes known as the Global South and Global North, respectively. Other terms used to define these countries include, low/middle/high income, Developing/Developed, two-thirds World/one-third World, and First World/Third World. Usage of the terms Majority and Minority World reflects the proportion of the world as divided by population and land mass. Whilst any terms that distinguish countries in a binary fashion are problematic, at times it is necessary and useful. The terminology of Majority/Minority World problematises knowledge production with the question, '[to] what extent is this "general knowledge" at all, given that the contexts on which it is built, are actually in a global minority?' (Jakobsen, 2012, p. 112). These terms serve as a discursive reminder that those in the Majority World do indeed encompass most of the world, in population and land mass. There is, however, a risk of oversimplification with categorisation of countries.

For example, countries such as China or Brazil do not fit neatly within any of these categories. Although I have defined Majority/Minority World here, throughout the book I have aimed to be specific when discussing countries, rather than relying on homogenising terms.

At times, I also use the term 'Africa' to denote the continent of Africa and the associated imagery and stereotypes held mainly by Minority World countries. The broad label 'Africa' is used by media, celebrities, politicians, and just about everyone; however, given there are 54 countries that make up the continent of 'Africa,' this label is often too broad and stereotypical – particularly given the associated imagery identified above. In adopting this broad term, I recognise the diverse countries and populations within the continent of Africa and recognise they represent a multiplicity of cultures and norms. Where possible, I have specified which African nation I am referring to, as to not proliferate the continuation of a broad term that does not recognise the diversity of African nations.

**Introduction to the research**

This book crosses the fields of education and tourism to explore tourism in one school in Matabeleland North Zimbabwe. It is based on ethnographic research of Matopo School that I conducted in 2019. The research project involved observations of the school for the duration of Term 1. The data collection included interviews with teachers, school founders and tourism personnel involved with the school. It also included fieldnotes (see Harper & Smithers, 2023) and art-based interviews with children in Grade 5 (see Smithers, 2022). All names of people, institutions, and other places in this book are pseudonyms, with all adults and children selecting their choice of pseudonym.

**Structure of the book**

In Chapter 2 I outline the entwined economic and educational histories of Zimbabwe. It begins by outlining a brief overview of Zimbabwe's economic history. Zimbabwe's economic system has experienced long periods of instability, so an understanding of the volatile economy is important. Next, Chapter 2 explores the current education challenges facing Zimbabwe and the previous history of education. The education system in Zimbabwe began as a relatively well-funded and well-resourced system; however, declines in the economy and political instability have severely influenced the ability for Zimbabwe to provide a functioning education system. These two sections set the scene for the rest of the book, as the economic and education history are fundamental to the development of a need for tourism to fund schools. With calls for the Zimbabwean tourism sector to move beyond a dominance of wildlife tourism towards a more 'culture' orientated tourism offering, in Chapter 2 I also explore the development of tourism as an economic salve for both schools and local communities in a time of dire economic crisis.

In Chapter 3 I provide an in-depth description of Matopo School – the focus of this book. I outline the involvement of the three main companies who fund the school, Education Tours, Adventure Company, and Shumba Safaris. The three companies have varying levels of engagement and provide in return varying levels of funding, which is explained. I also briefly outline my role in the school and the way the research was shaped by the school, as well as exploring my own whiteness and privilege and how it relates to the research.

In Chapter 4 I outline the conceptual underpinnings of the book. Using development and tourism, I discuss dominant discourses that frame the school tour and explore these in relation to the metaphor established in this introductory chapter. I argue that there are discourses which frame the implementation of the tourism and discuss the structural influences of these discourses. First, I provide a brief history of Development to inform the discussion in later chapters. I provide an insight into tourisms that are based in development and argue that school tours are an activity that occurs across a range of niche tourisms. Then, using school visits as part of volunteer tourism as a guiding example, I identify how two particular imageries shape the conditions of possibility for tourisms guided by development. The first, the white saviour complex, is explored through a discussion of celebrity humanitarianism and discourses of saving the world. The second, the poor-but-happy-child is explored through a discussion of volunteer tourism and dominant images of Majority World children in the Minority World. This chapter lays the groundwork for the remaining chapters in exploring the discursive structures which frame the school tour.

The metaphor – of a movie with its actors, script, and set design – that opened this chapter is used to frame this book. These representations form a fusion of images which circulate around stereotypes of Africa, ensuring tourists arrive with preconceived notions of what they will see in 'Africa'. Essentially, tourists know exactly how the movie will end because they have seen it, in various forms, many times. There are a set of expectations which comprise the shape of the script, the movie's set design and the actions of those in the scene. The empirical chapters are structured according to the key 'players' in the metaphor. These key players include the tourism personnel, teachers, and school children.

In the first of these chapters (Chapter 5) I focus on the school teachers and their role in the production and maintenance of the 'image of Africa'. Firstly, I argue that the school tour constructs and maintains a heterotopia of the image of Africa, and that the teachers play a fundamental role in the continuation of this image. Second, I discuss the conflicting roles of the school teachers as they balance the need to present an Image of Africa, and also to perform their role as effective classroom teacher. This chapter concludes by exploring why the teachers prioritize the use of children during the school tour, despite the disruption to the children's education. Children's art, drawn by Grade 5 students, is used in this chapter.

In Chapter 6 I examine the 'starring role' of the children in the metaphor and outline the ways in which they are expected to conform to the image of the 'poor-but-happy' child. I begin by identifying how the role of the 'poor-but-happy child' is constructed in a particular way, as to allow the tourists to feel a level of comfort at the poverty they witness at the school. The second section outlines the 'love' the tourists are reported to have for the children, its discursive basis, and the reaction of the teachers to the way their relationship to the children is constituted. The teachers' opinions are important here as they are the directors of the film in which the poor-but-happy children are the main characters. The third section examines the naming of the tourists and argues that they are discursively positioned as 'friends', or 'visitors' rather than tourists, as to help to 'sell' the image of Africa and encourage participation and donations.

In Chapter 7 I explore the role of the white saviour (tourist) in the metaphor. I outline how the needs of the school are often not considered, rather the tourists anticipate what they see as the needs of the school. This chapter begins by exploring the problematic white saviour role in depth, particularly in relation to gift-giving practices the tourists enacted whilst in the school. Second, this chapter explores the role of whiteness and the white saviour role in Development more broadly through a discussion of one tension involved, that of having external funders financing the school. Last, this chapter explores whiteness in Zimbabwe through a discussion of the relationship of the local community to the school founders. It is in this chapter that I play with the concept of dilemmas of interpretation in understanding the context of the local white community and the school.

In the final chapter I bring together the conclusions of the book to identify the global and local repercussions of the school tour. A way forward is presented through the voices of the teachers, school founders, and the tourism personnel. This includes practical solutions that are grounded in the suggestions of those who work in the school on a day-to-day basis. These solutions are explored in light of the existing literature about school tours.

## Note

1  This is referenced by Johnson as translated (by Johnson) from a recording of a radio interview which has not been published in English.

## References

Abraham, R. (2003). The localization of 'O' level art examinations in Zimbabwe. *Studies in Art Education*, *45*(1), 73–87.

Ahmed, S. (2007). A phenomenology of whiteness. *Feminist Theory*, *8*(2), 149–168. https://doi.org/10.1177/1464700107078139

Baptista, J. A. (2011). The tourists of developmentourism– representations 'from below'. *Current Issues in Tourism*, *14*(7), 651–667. https://doi.org/10.1080/13683500. 2010.540314

Brandt, F., & Josefsson, J. (2017). Sexuality and power on South African game farms; reflections on positionality and emotions in ethnographic research. *Emotion, Space and Society*, *23*, 26–32. https://doi.org/10.1016/j.emospa.2017.02.004

Chitiyo, G., Chitiyo, M., Rumano, M., Ametepee, L. K., & Chitiyo, J. (2010). Zimbabwe education system: Emerging challenges and the implications for policy and research. *Journal of Global Intelligence and Policy*, *3*(3), 35–42.

Crossley, É. (2012). Poor but Happy: Volunteer Tourists' Encounters with Poverty. *Tourism Geographies*, *14*(2), 235–253. https://doi.org/10.1080/14616688.2011.611165

Dekeza, C. (2018). *The responsiveness of rural primary schools to the needs of orphaned and vulnerable children in Zimbabwe* University of South Africa]. http://uir.unisa.ac.za/bitstream/handle/10500/25059/thesis_dekeza_c.pdf?sequence=1&isAllowed=y

Durrheim, K., Mtose, X., & Brown, L. (2011). *Race trouble: Race, identity and inequality in post-apartheid South Africa*. Lexington Books.

Dzimiri, P., & Runhare, T. (2012). The human security implications of Operation Restore Order on urban habitation in Zimbabwe. *Journal of Human Ecology*, *38*, 191–205.

Fanon, F. (1986). *Black skin, white masks*. Pluto Press.

Foucault, M. (1967/1984). Des espace autres [Of other spaces: Utopias and heterotopias]. *Diacritics*, *16*(1), 22–27.

Gardner, B. (2016). *Selling the Serengeti: The cultural politics of safari tourism*. The University of Georgia Press.

Garland, E. (2008). The elephant in the room: Confronting the colonial character of wildlife conservation in Africa. *African Studies Review*, *51*(3), 51–74. https://doi.org/10.1353/arw.0.0095

Garland, E. (2012). How should anthropologists be thinking about volunteer tourism? *Practicing Anthropology*, *34*(3), 5–9.

Harper, M., & Smithers, K. (2023). Fieldwork from A–Z? Exploring shifting identities in doctoral research in Australia and Zimbabwe. *The Australian Educational Researcher*. https://doi.org/10.1007/s13384-023-00649-1

Haslanger, S. (2019). Tracing the sociopolitical reality of race. In J. Glasgow, S. Haslanger, C. Jeffers, & Q. Spencer (Eds.), *What is race? Four philosophical views* (pp. 6–37). Oxford Scholarship Online. https://doi.org/10.1093/oso/9780190610173.003.0002

Hays, C. M. (2012). Placing nature(s) on Safari. *Tourist Studies*, *12*(3), 250–267. https://doi.org/10.1177/1468797612461086

Hlupo, T., & Tsikira, T. (2012). Comparative analysis performance of satellite primary schools and their mother schools in Masvingo Province, Zimbabwe. *Journal of Emerging Trends in Educational Research and Policy Studies*, *3*(5), 604–610. https://hdl.handle.net/10520/EJC127701

Hong, S., & Vicdan, H. (2016). Re-imagining the utopian: Transformation of a sustainable lifestyle in ecovillages. *Journal of Business Research*, *69*(1), 120–136. https://doi.org/10.1016/j.jbusres.2015.07.026

hooks, B. (1992). *Black looks: Race and representation*. South End.

Jakobsen, H. (2012). Focus groups and methodological rigour outside the minority world: Making the method work to its strengths in Tanzania. *Qualitative Research*, *12*(2), 111–130. https://doi.org/10.1177/1468794111416145

Johnson, P. (2013). The geographies of heterotopia. *Geography Compass*, *7*(11), 790–803. https://doi.org/10.1111/gec3.12079

Jones, J. L. (2010). 'Nothing is straight in Zimbabwe': The rise of the kukiya-kiya economy 2000–2008. *Journal of Southern African Studies*, *36*(2), 285–299. https://doi.org/10.1080/03057070.2010.485784

Landau, P. S. (2002). An amazing distance: Pictures and people in Africa. In P. S. Landau & D. D. Kaspin (Eds.), *Images and empires: Visuality in colonial and postcolonial Africa* (pp. 1–40). University of California Press.

Madziyire, G. T. (2015). *Evaluating the impact of philanthropic activities in public high schools in Mutasa district, Zimbabwe: An educational management perspective* University of South Africa]. https://core.ac.uk/download/pdf/43177311.pdf

Magudu, S. (2020). Configuring the key social justice concerns in rural education in Zimbabwe. In A. Masinire & T. M. Ndofirepi (Eds.), *Rurality, social justice and education in Sub-Saharan Africa Volume 1: Theory and practice in schools* (pp. 63–86). Palgrave MacMillan.

Mangena, T., & Ndlovu, S. (2013). Decolonization-oriented naming of selected 'satellite' schools in Zimbabwe. ICONN 2.

Mangwaya, E., Jeko, I., & Manyumwa, C. (2013). Availability of print curriculum materials and its consequences for the quality of education in schools located on newly resettled farm areas in Zimbabwe. *Asian Social Science*, *9*(1), 249–256.

Martins, C. (2011). The dangers of the single story: Child-soldiers in literary fiction and film. *Childhood*, *18*(4), 434–446. https://doi.org/10.1177/0907568211400102

Masschelein, J. (2018). An educational cave story (on animals that go to 'school'). In P. Smeyers (Ed.), *International Handbook of Philosophy of Education* (pp. 1185–1200).

McClintock, A. (1995). *Imperial leather: Race, gender and sexuality in the colonial contest.* Routledge, Inc.

Mudimbe, V. Y. (1988). *The invention of Africa: Gnosis, philosophy and the order of knowledge.* Indiana University Press.

Mudzingwa, K., & Magudu, S. (2013). Idealism versus realism: Expectations and challenges of beginning teachers in three districts of Masvingo Province, Zimbabwe. *Journal of Studies in Social Sciences*, *3*(1), 33–54.

Riddell, A. (1998). Book review: Policy, evaluation and leadership: The context of educational change in Zimbabwe. *Compare: A Journal of Comparative and International Education*, *28*(1), 115–122. https://doi.org/10.1080/0305792980280110

Rideout, L. (2011). Representations of the 'Third World' in NGO advertising: Practicalities, colonial discourse and western understandings of development. *Journal of African Media Studies*, *3*(2), 25–41. https://doi.org/10.1386/jams.3.1.25_1

Sheridan, V., Landy, D., & Stout, V. (2019). The return of the 'Black Babies': How development education affects schoolchildren's attitudes to the majority world - the Trócaire 2012 Lenten campaign. *Race, Ethnicity and Education*, *22*(6), 858–874. https://doi.org/10.1080/1361332.2017.1417829

Sithole, F. (2017). *Strategies to enhance teacher motivation in satellite secondary schools in the resettlement areas of Matabeleland North province* Zimbabwe Open University]. https://www.academia.edu/67625209/Strategies_to_enhance_teacher_ motivation_in_satellite_secondary_schools_in_the_resettlement_areas_of_ Matabeleland_North_Province

Smithers, K. (2022). 'What is your name, where do you come from, what is your grade?' Using art-based interviews to highlight the experience of children hosting school tours in Matabeleland North, Zimbabwe. *Tourism Recreation Research*, *48*(6), 912–924. https://doi.org/10.1080/02508281.2022.2133812

Smithers, K. (2023). Heterotopia and the 'image of Africa': School tours and philanthropy in a Zimbabwean school. *Critical Studies in Education*, 1–17. https://doi.org/ 10.1080/17508487.2023.2286230

Smithers, K., & Ailwood, J. (2022). Development tourism and school tours in Zimbabwe. In M. Novelli, J. M. Cheer, C. Dolezal, A. Jones, & C. Milano (Eds.), *Handbook of Niche Tourism* (pp. 345–356). Edward Elgar.

Staples, A. J. (2006). Safari adventure: Forgotten cinematic journeys in Africa. *Film History: An International Journal*, *18*(4), 392–611.

Tarisayi, K. S. (2019). A school in distress: The manifestations of poverty at a selected satellite school in Masvingo district, Zimbabwe. *Journal of Geography Education for Southern Africa*, *4*(2), 79–96.

Tarisayi, K. S., & Manik, S. (2019). The role of land reform beneficiaries and the reasons for them developing and supporting a satellite school in Masvingo, Zimbabwe: A social capital marriage of Nhimbe and allied reasons. *Education as Change, 23*.

Utete-Masango, S. J. (2016). *Education sector strategic plan 2016–2020*. Online: https://www.globalpartnership.org/node/document/download?file=document/file/2016-07-education-sector-strategic-plan.pdf

# 2 How does tourism come to be in a school?

I arrive at the meeting point at 7am, a busy spot on the main road out of town. Everybody knows this spot; it is a hub of informal transportation and ridesharing. It is early morning but already hot, humid and dusty. I find one of the teachers from Matopo School and we greet each other while agreeing it is far too hot already. It is the wet season but we are in an unseasonable drought. Once the other teachers arrive, we wait to see if our taxi will appear. While we wait, I consider the scene before me. Safari vehicles pass by, either heading out of town filled with tourists in khaki on their way to spot animals or filled with workers heading into town for a day of work. There are also air-conditioned minibuses transporting tourists in and out of town, and I know these minibuses will contain eskies of ice-cold bottled water – to ensure passenger comfort. Around us, at the meeting point there is a bustle of activity as people negotiate how much they will pay for a seat in a car heading out of town – a very common practice. There are many people already here, at least 20. Nearby, there is a vendor selling cool drinks and phone credit. This spot is an example of the informal economy in Zimbabwe, in which people make informal sales based in cash transactions.

We are waiting for our taxi, a pre-arranged ride who will take us to school, although sometimes he arrives late or not at all. If he does not arrive it is most likely that his car has had some sort of mechanical failure or he is in a long line to buy fuel. The area of Zimbabwe I am in suffers from fuel shortages, and I often see lines of cars waiting for the fuel tanker to arrive. When there is fuel, it is very expensive and during my stay in Zimbabwe there are headlines about Zimbabwe having the highest priced fuel in the world. Those who do have cars help to alleviate costs by charging hitchhikers for a seat in their car. This morning, if our taxi does not arrive, we too will join the negotiations of finding a car with space that can take us as far as the turnoff road to our school. If the taxi does not arrive, we will be late and we will need to walk several kilometres from the main road to school. In the afternoon, we repeat the process, as we wait to see if our taxi will arrive. If it does not arrive, we will walk to the main road and hitchhike back into town. Each time the taxi does not arrive we must negotiate with people on a price for a ride, trusting strangers to drive us as far as we need to go – often without seatbelts and in an overloaded vehicle.

DOI: 10.4324/9781003407980-2

For many teachers in Zimbabwe, the challenge of their workday begins before they arrive on the school grounds. Fuel shortages, inflation, and wealth disparities mean most teachers in rural schools must either live on-site in teacher accommodation (mostly without power or flushing toilets) or live further away and hitchhike, negotiate a ride, or walk to school. Those who do not have access to on-site accommodation or those who need to live with their families must constantly navigate the changing economic environment and how this affects their travel to school, which includes inflation of the local currency and hard to find necessities. The challenge of transport is not unique to teachers. During my time at Matopo School I met a policeman who worked a few towns away and would also pick up passengers at the meeting spot to fund the fuel costs associated with working so far away.

The challenges of travel to school are just one component of the challenges that teachers in Zimbabwe endure as part of their everyday experience. Schools are also under-resourced and teachers' salaries are often unpaid by the government, or do not match the ever-increasing inflation. For the schooling system in Zimbabwe, there is an intimate connection to the success and failures of the economy. The twin histories of the Zimbabwean economy and the schooling system illuminate the ways in which there are no simple solutions to the problem of resourcing schools in Zimbabwe, beyond significant restoration of the economy in Zimbabwe.

### Economic challenges: a brief history

Zimbabwe has experienced one of the most severe hyperinflation events in this century. At the peak, the daily inflation rate was roughly 98%, with prices doubling every 24 hours (Hanke & Krus, 2013). The economic history makes it both an interesting and challenging place to live, work, and research; forever present is the spectre of previous economic collapse and future economic troubles.

Zimbabwe had a decade of relative economic success following independence in 1980 (Lines, 1988; Masaka, 2013; Meisenhelder, 1994). Some credit this to the initial socialist-leaning policies which included large public investments in education and other infrastructure (Dorsey, 1989; Kanyongo et al., 2006). These socialist policies were cut short in the early 1990s as Zimbabwe moved towards economic support from the World Bank. The World Bank pushed structural adjustment programmes that included market-economy and neoliberal ideologies with an emphasis on cuts to public spending (Kawere & Dibie, 2000). It is worth noting here that interventions, such as those from the World Bank and the International Monetary Fund (IMF), have led to large levels of debt and dependency, in both Zimbabwe and more broadly amongst many low to middle income countries (Murithi, 2011).

The introduction of the Economic Structural Adjustment Programme (ESAP) in 1991, Zimbabwe's military involvement in the Democratic Republic of Congo, and land reform policies created an environment for sharp economic decline in the late 1990s (Coomer & Gstraunthaler, 2011; Masaka, 2013). One-off payments to Independence war veterans and involvement in the Democratic

Republic of the Congo in 1997 caused foreign investors to panic about economic instability and withdraw their support and funding for Zimbabwe (Mlambo, 2017). Further, the earlier introduction of ESAP in 1991 meant a reduction in public spending, as ESAP had a particular focus on a free-market economy and cuts to public investment in schooling (Abraham, 2003; Johnson, 1990). These decisions, coupled with continued poor management of the economy created further budget deficits over the following years (Coomer & Gstraunthaler, 2011; Masaka, 2013). As budget deficits increased, government printing of money also rose, and Zimbabwe entered a period of extreme hyperinflation in 2006–2008; in July 2008 the inflation rate was 231,150,888.87% which rose to 89,700,000,000,000,000,000,000% by mid-November 2008 (Dekker, 2017). The period of hyperinflation included the printing of bank notes with a face value of up to 100 trillion Zimbabwean dollars.

The economic decline was partially stalled with the election of a coalition government between 2008 and 2013, and the introduction of US dollars which created a more stable multi-currency economy (Dekker, 2017; Noyes, 2020). At this time, the Zimbabwean dollar was largely abandoned and was eventually de-monetised in 2015. In more recent times, these bank notes are sometimes sold by street vendors to tourists as a souvenir. In the coalition period, the economy seemed to stabilise; however, after 2013 the economy continued to falter. More recently, bond and treasury notes were introduced in further attempts to reduce inflation with the aim of controlling how much currency could circulate at any one time (Southall, 2018). The introduction of bond and treasury notes have had very little effect, as there is a generalised mistrust of government interventions in the economy, and they were largely perceived as an attempt to reintroduce the Zimbabwean dollar; as a result inflation has continued to rise since 2017 (Noyes, 2020; Southall, 2018). On February 20, 2019, Zimbabwe introduced another form of currency, the Real Time Gross Settlement dollar (RTGS$) in a move to de-dollarise the economy (Imam, 2022). The attempts to de-dollarise have created a black market of currency trading in which bond notes are traded for US dollars at a reduced rate.

In 2019, the time I was there, the annual inflation rate for Zimbabwe was 521% (Zimbabwe Tourism Authority, 2019). The same year, petrol prices became the most expensive in the world, and steadily rose from 3.34RTGS$ in January to 18.03RTGS$ in December (ZIMSTAT, 2019).[1] Despite recent efforts to stabilise the economy, it is clear that Zimbabwe is still struggling to recover from the economic decisions made since the early 1990s and into the 2000s.

Despite economic turbulence, Zimbabwe appears to be doing relatively well in terms of income distribution. The Gini coefficient is a measure of equality in an area or country, in which zero represents complete income equality and one representing that one person holds the total wealth of the country. The most recent Gini coefficient for Zimbabwe was 0.44 in 2017 (ZIMSTAT, 2017b). In terms of Zimbabwe's bordering countries, according to the World Bank (World Bank, 2020a), the Gini coefficient is 0.63 in South Africa (2014), 0.57 in Zambia (2015), 0.59 in Namibia (2015), 0.54 Mozambique (2014), and

0.53 in Botswana (2015). Despite relative income equality there is a rural/urban divide, with low-income families disproportionately represented in rural areas with 90% of households considered 'poor' residing in non-urban areas (ZIMSTAT, 2017b). This is evident in the distribution of toilets, with 91.5% of urban households having a flushing toilet and 36.8% of rural households having no toilet at all (ZIMSTAT, 2017b).

There is another layer of complexity to the distribution of aid and cash flows across the borders. After the economic turbulence of 2000–2008, many Zimbabweans left the country and in the present, many Zimbabweans who cannot find work within Zimbabwe may relocate legally, or illegally, to neighbouring countries to find work (Chitiyo & Chitiyo, 2009). Usually, those who leave are expected to send remittances to family living in Zimbabwe (McGregor & Pasura, 2010). These remittances increase informal cashflows across borders. As a result of migration, children are left behind with grandparents when their parents leave for work in other countries, or with other relatives (Makaudze, 2013). In practice, the remittances are not always provided to families who take on caregiving roles. For schools, this has implications as families may take on extra children to support their extended family but may not be able to pay the school fees and other associated school costs for these children. The functioning of the economy has a flow on effect to schooling and school systems – beyond just simple resourcing.

### Education challenges: a system in crisis

The structure of Zimbabwean schooling is very similar to other countries. In Zimbabwe, students undertake four years of 'infant education' which comprises of two years in Early Childhood Development and two years of early primary education (Grade 1 and Grade 2). This is followed by five years of 'junior education' consisting of Grades 3 to 7. After completion of Grade 7 examinations, students complete four years of lower secondary school which concludes with O-level examinations. Some students will then proceed to two years of upper secondary education (A-level), polytechnic college, or teachers' college. Upper secondary school is not compulsory, and students must pass four O-level classes for entry.

Zimbabwe has a somewhat successful, but under-resourced and inequitable, education system:

> No doubt, in a developing country like Zimbabwe and indeed the entire continent of Africa, education becomes a fundamental ingredient in the recipe for development. Education can thus be used as a barometer to measure and project a country's overall performance in terms of human development. Unfortunately, having started off so well in the 1980s, Zimbabwe's education system will probably receive a fail grade for the past decade – a result for which the country will pay heavily in the foreseeable future.
>
> (Chitiyo et al., 2010, p. 39)

Zimbabwe's education system is considered by many to be in the midst of a crisis (e.g. Chitiyo et al., 2010; Chitiyo, 2014; Hove & Ndawana, 2019; Tarisayi, 2019; Tarisayi & Manik, 2019; Tawodzera & Themane, 2019). The crisis is linked to both the current economic climate and economic decisions made in the late 1990s.

The Zimbabwean school system has three school types: registered, satellite, and unregistered. Schools can become registered when they meet certain infra-structure requirements, including a certain number of teacher housing, class-rooms, and facilities. In terms of authorities responsible for schools there are two broad categories: government, which make up 87.8% of primary schools, and non-government (Ministry of Primary and Secondary Education, 2019). Government schools have several classifications: government, other-government line ministries, city council, district council, and town board (Ministry of Primary and Secondary Education, 2019). The non-government classifications are church/mission, mine, private company, farm, trust, private individual, and other. In Matabeleland North, there are a range of non-government authorities responsible for schools, such as churches, farms, mines, private companies, and trusts. For the purposes of anonymity, the school in this study is a satellite school run by a non-government authority that will not be named.

The literacy level in Zimbabwe is 94% overall, with older people (over age of 55) experiencing lower levels of literacy (ZIMSTAT, 2017a). The total number of primary school aged children in 2019 was 2,585,626, of which 96.4% were enrolled in school (Ministry of Primary and Secondary Education, 2020). In Matabeleland North, the province this research was conducted in, there were 166,518 primary school aged children enrolled in 2019 (Ministry of Primary and Secondary Education, 2020). Across Zimbabwe in 2019, the completion rate for primary school was 75.65% and lower secondary completion rate was 71.8%; however, the figures are lower for Matabeleland North at 74.76% and 62.0%, respectively (Ministry of Primary and Secondary Education, 2020). A-level completion rates are low. Across Zimbabwe, 33% of secondary schools offer A-level classes and only 10% of students who finish Grade 7 will remain enrolled by their final year of school (ZIMSTAT, 2014). There is an urban/rural divide, with the requirement to pass four O-level classes, coupled with under-resourced schools who may not provide A-level classes to continue into, limiting the ability of rural students to enter university (Kanyongo, 2005).

The current struggles faced by the education system are not new. They have a far-reaching history that begun in the late 1980s but were sparked by the design of the colonial education system. During colonial administration, it was mostly white middle-class children who proceeded through schooling due to costly school fees and strict admission requirements (Abraham, 2003). As only a relatively small number of children attended school, most schools were private and well-resourced (Abraham, 2003). If non-white children did attend school, they were placed into a limited vocational stream unless they were in the highest achieving band of students (Dorsey, 1989; Mungazi, 1985).

The number of students who progressed through schooling is contested, with some saying it was 20% of the Grade 7 pupils who passed the examination for entry into lower secondary school (Chitiyo et al., 2010) and others who suggest the number was 13% (Mutema, 2014); however, actual secondary school enrolments represented less than 10% of the population (World Bank, 2020c).

Following independence in 1980, the government introduced policies with an aim of education becoming free and open for all students. Some have linked this policy to the socialist economic practices of the Zimbabwean government in the 1980s, as discussed above (Abraham, 2003). Due to the policy change, there was an influx of students including some children who were older than primary school age. Prior to independence, only 60% of primary aged children were enrolled in primary school. Following independence, enrolment numbers exceeded the total number of children of school going age in Zimbabwe, as children who had previously been excluded were able to enrol in primary school, regardless of their age or prior achievement (World Bank, 2020b).

The policy of free and open education meant that a very large number of students enrolled in an education system designed for roughly 60% of the primary school age population. These policy changes following independence, although fair and equitable, created many problems (Chitiyo & Chitiyo, 2009). In the first decade of Zimbabwe's independence the number of primary schools increased by 88%, from 2,401 to 4,504 in 1989 (Ministry of Primary and Secondary Education, 2019). The number of primary schools was 6,288 in 2018 (Ministry of Primary and Secondary Education, 2019). The change from mostly private, selective schools to free and compulsory education placed pressure on existing school infrastructure, such as desk numbers and textbook availability and upon teacher colleges to produce teachers quickly (Madziyire, 2015; Riddell, 1998). Coupled with the economic collapse, there are continued problems for the education system today. For example, many schools use 'hotsitting', in which half the school attend in the morning and the other half in the afternoon; or one site may serve both a primary and secondary school in this manner (Dzimiri & Runhare, 2012).

As detailed, the primary school system was not prepared for such an increase, having previously served the very small white population (Abraham, 2003). Whilst the existing primary school infrastructure struggled to acclimatise to the rise in student enrolments; the secondary system had a disastrous transition, mainly due to a lack of infrastructure and teachers available (Chitiyo et al., 2010). Previously, primary schools had catered to roughly 60% of the population, whilst secondary schools had been set up to cater to an even smaller minority white population. Coupled with an education policy of 'open to all', another policy meant all Grade 7 students would proceed to lower secondary schooling, whether they passed the entrance exam or not (Ministry of Primary and Secondary Education, 2019). The number of secondary schools was 177 in 1979 and rose to 1502 by 1989, with this increase putting great pressure on the secondary school system in terms of staffing and resources (Ministry of Primary and Secondary Education, 2019). In the present, the secondary school system could be considered to be in crisis based on student outcomes alone; in Matabeleland North only 27.5% of students passed Grade 7 examinations in 2019, with 47% of students passing nationally (ZIMSTAT, 2020); however, Bulawayo and Harare, as urban centres, both had pass rates of over 72% in 2019 (ZIMSTAT, 2020). This highlights the unequal achievement rates between urban and rural areas, which continues in the present.

It is important to note that Zimbabwe has faced several issues in terms of school continuation and consistency. Firstly, there has been a devaluation of all levels of education in Zimbabwe, meaning that there are very few employment opportunities, even for qualified graduates (Makaudze, 2013). Secondly, the schooling system faced significant disruptions in both 2008 and in 2020. In 2008, some schools were closed for the entire year, as teachers' salaries were unpaid or completely devalued due to inflation (Campbell et al., 2014). During the economic turbulence of 2008 and the years before, there was also significant brain drain as qualified teachers left to pursue teaching careers in other countries (Crush & Tawodzera, 2013). There were many teachers I met during my time in Zimbabwe who had spent time teaching in South Africa. In 2020, the COVID-19 pandemic generated another year of school closures (ZIMSTAT, 2020). This, again, disproportionately affected rural children with only 25% engaging in any sort of education or learning when their schools closed and only 6% of students having any contact with teachers, compared to 70% and 40% respectively for urban students (ZIMSTAT, 2020).

There have been reports that teachers are underqualified due to the 'brain drain' that occurred prior to and immediately following the economic collapse in 2008 (Mlambo, 2017). Recently, there has been a concerted effort to increase the number of qualified teachers; in 2018, 97.14% of primary teachers and 85.47% of secondary teachers held some form of teaching qualification (Ministry of Primary and Secondary Education, 2019). In Matabeleland North, most teachers have a Diploma or Certificate in Education, few have a Bachelor's degree (Ministry of Primary and Secondary Education, 2017). Whilst most teachers have some form of qualification, there is a reliance on textbook instruction; with an established lack of textbooks available, this has ramifications in terms of student learning as teachers may have limited access to textbooks (Madondo, 2020; Madziyire, 2015).

In the present, to compensate for a lack of government funding, schools often charge fees to pay teacher salaries and to purchase resources such as textbooks. Most schools face shortages of furniture, other infrastructure and curriculum materials, particularly in rural areas (Magudu, 2020; Mangwaya et al., 2013). Schools are understaffed and under-resourced in terms of classrooms and classroom teachers, with the average pupil to classroom ratio 45:1 (Utete-Masango, 2016). While these resources – such as infrastructure, teachers and textbooks – should be provided by the government in government schools, they are often inadequate or not provided, with most of the government budget for education allocated to teacher salaries (Hlupo & Tsikira, 2012). Despite the budget allocations, government teacher salaries, if paid at all, are just on the poverty line at roughly US$6000 a year (Utete-Masango, 2016). Coupled with the continued hyperinflation, these salaries are increasingly unable to provide teachers with liveable salaries.

Satellite schools have become a necessary component of the Zimbabwean system, as they help to improve the resourcing and facilities of schools in Zimbabwe. Satellite schools were introduced as a temporary solution to the low number of schools in land reform areas, but have now become a permanent feature of the Zimbabwe school system as other communities develop schools on communal land (Mudzingwa & Magudu, 2013; Tarisayi, 2019). Satellite schools are usually built and funded by the community, with the government sometimes providing teachers to the community (Tarisayi & Manik, 2019). The Ministry of Primary and Secondary Education (MOPSE) may appoint a Teacher in Charge, who is a government employee and responsible for administration. Although not government schools, satellite schools must follow MOPSE curriculum and policies. Further, each satellite school is partnered with a registered school, with Matabeleland North having 135 satellite schools and 479 registered schools in 2018 (Ministry of Primary and Secondary Education, 2019). Satellite schools are not officially registered with MOPSE, which means children cannot sit their Grade 7 examinations at their home school; instead they must travel to the 'parent' school to sit examinations (Tarisayi, 2019; Tarisayi & Manik, 2019). What began as a temporary solution is now a very permanent fixture in the Zimbabwean education system.

Satellite schools can become registered with the government when they meet minimum standards for hygiene and school infrastructure such as a minimum number of toilets or desks. Satellite schools usually have very little permanent infrastructure and struggle to meet the minimum standards for registration set by MOPSE (Ministry of Primary and Secondary Education, 2019). As satellite schools' infrastructure and resources are not funded by the government (Mangena & Ndlovu, 2013), some hope to receive donations from donors and NGOs (Magudu, 2020; Sithole, 2017; Tarisayi, 2019). The school in this study was a satellite school in the process of gaining official registration and two of the ten teachers were employed by MOPSE, although the school supplements their government salary through extra payments in US dollars.

Many schools in Zimbabwe use philanthropic donations as a source of infrastructure and funding (Madziyire, 2015). In fact, 'visible philanthropy in education in Zimbabwe seems to come largely from foreign organisations – so much so that in Zimbabwe the term "donor" is synonymous with assistance from foreign-based organisations' (Madziyire, 2015, p. 2). Some schools, especially independent, satellite, and/or community-based schools, encourage foreign donors to sponsor individual students (Dekeza, 2018). Madziyire (2015) found that donations of school fees for vulnerable students was an important factor to prevent students dropping out of school.

There are many types of organisations involved in philanthropic funding of schools in Zimbabwe. As part of corporate social responsibility activities, small and medium enterprises donate school fees and classroom infrastructure

(Chanakira & Masunda, 2019). Charities developed by the Zimbabwean diaspora are also involved in schooling (Mutambasere, 2022). In areas where mining occurs, there are schools funded by mining companies or other schools who receive stationery donations from mining companies (Nhavira, 2019). Tourism providers often partner with charitable organisations, such as the Grand Circle Foundation, to provide school infrastructure or scholarships (Novelli, 2016; Snyman & Spenceley, 2019).

With the rise of tourism-oriented NGOs in Zimbabwe, schools have found other avenues for funding and become more resourceful in the manner in which they advertise themselves. One such avenue is developmentourism, as the school does not need to make time to find tourists, the tourists are brought to them through mass tour packages. During such tours, schools encourage tourists to sponsor their students, which brings developmentourism into the home-sphere, not just the school yard.

Whilst this brief overview potentially paints a sad state of affairs, there are many teachers and students in Zimbabwe who attend school each day in the hopes of a different future. Some have described Zimbabwe as a phoenix rising from the ashes of the legacy of colonisation (Chung, 2014). Schools are showing ingenuity in their response to the ongoing under-resourcing of schools and economic challenges. Some have referred to this 'cleverness, dodging, and the exploitation of whatever resources are at hand, all with an eye to self-sustenance' as *kukiya-kiya* (Jones, 2010, p. 286). In consideration of the statistics described here, it is important to remember that these statistics are Minority World measures of progression and may not reflect the shared resources that are spread across a community. The school in this study is an example of the ways in which Zimbabweans are enacting *kuronga-ronga* (make a plan) towards improving education in Zimbabwe (Jones, 2010). Despite the challenges the education system faces, education is held with a high regard in Zimbabwe.

**Tourism in Zimbabwe**

Although seemingly disparate areas of study, through developmentourism, tourism is being brought into the educational sphere. The school tour is a tourism venture, which takes place in a school. Having detailed the economic history and the education system which make this type of tourism possible, I now explain the current state of tourism in Zimbabwe as a means to explain the particular discourses privileged in representations of the generic 'Africa'.

Africa, as a continent, has a 5% share of tourist arrivals globally (Zimbabwe Tourism Authority, 2019). Zimbabwe receives roughly between 1.8 million and 2.5 million tourist arrivals per year (Trading Economics, 2020). In 2019, there was a reduction of tourist arrivals in Zimbabwe by 11%, perhaps due to a rising negative destination image caused by the turbulent economy, power cuts, and fuel shortages (Zimbabwe Tourism Authority, 2019). Tourism injects around US$1.3billion into the economy in Zimbabwe and most of this, 70%, is from foreign tourist arrivals rather than domestic tourists (Zimbabwe Tourism

Authority, 2019). A large proportion of tourist arrivals is from other countries in Africa which represents in part, the level of mobility in the region. In terms of intercontinental travel, United Kingdom and United States tourists represent the largest foreign arrivals each year (Zimbabwe Tourism Authority, 2019).

The political and economic climate affects the tourism industry. Tourist guides in Zimbabwe are often underpaid, and there are inconsistencies in the training that leads towards qualifications (Nyahunzvi & Njerekai, 2013). In part due to the poor economy, tourism sites are often under-resourced and lack the infrastructure expected by tourists (Manwa, 2003; Ndlovu & Heath, 2013; Nyahunzvi & Njerekai, 2013). One strategy for the development of tourism has been the increased use of online marketing (Maswera et al., 2008). Volunteer tourism has relied on the image of Africa as a 'needy' country, with volunteer tourism companies advertising the need for Minority World tourists to 'come and make a real difference' (Nyahunzvi, 2013). Online marketing has helped Zimbabwe to increase its tourist numbers; however, the economic uncertainty and political instability have meant that some tourists consider Zambia to be a safer tourist destination (Govere et al., 2013; Nyahunzvi, 2013). A key feature in the potential growth of Zimbabwe's tourism industry is government involvement in initiatives that will encourage economic stability and promote tourism in Zimbabwe (Muzapu & Sibanda, 2016).

A suggested 'solution' to under-resourcing of the tourism industry has been community-based tourism (CBT). In Zimbabwe, the Communal Areas Management Programme for Indigenous Resources (CAMPFIRE) project was a community-based sustainable development programme initiated in 1989, in which communities would provide land for tourism opportunities and receive benefits such as a percentage of the gross income of tourism vendors and increased infrastructure in return. It began as a promising initiative, but management issues and a lack of reinvestment caused its ultimate decline in popularity by the early 2000s (Balint & Mashinya, 2006). In the present, CBT presents an opportunity for companies to perform Corporate Social Responsibility (CSR) that is giving back to the local community, whilst marketing themselves as a more ethical tourism provider (Chilufya et al., 2019). There are two approaches to CBT, top-down and bottom-up, with bottom-up approaches considered to be more beneficial for the community but with limited expansion opportunities (Zapata et al., 2011). There are three elements that are considered integral to CBT projects: located within community land, owned by the community, and managed by the community. The school tour in this study meets two of the requirements of CBT, it is located on community land and owned by the community, but not managed by the community.

There are mixed results for communities living near to tourism destinations in Zimbabwe. Tourism operators are often international companies or based in urban centres, which means rural residents often receive little benefit from the tourism in their area (Mandevani, 2014; Mutanga et al., 2015). In the area surrounding the Great Zimbabwe ruins in Masvingo province, traditional leaders are key informants and stakeholders in the development and implementation of tourism (Taru & Gukurume, 2013). The hotels in this area use local produce and labour, and sponsorship of local children by tourism vendors furthers the positive relationship with the community. In contrast, in Bulawayo residents believe that tourism does not benefit their community; instead tourism vendors and government officials benefit (Moyo & Tichaawa, 2017). In Victoria Falls, residents believed that tourism had a positive effect but raised prices of houses and goods (Tichaawa & Mhlanga, 2015). Further to this, respondents in a study by Mudimba and Tichaawa (2017) reported that they liked tourism in Victoria Falls because it provided employment in a jobless economy.

The main drawcard for tourists to Zimbabwe is the animal safari industry in which tourists can see the 'big five': buffalo, elephant, rhinoceros, leopard, and lion (Manwa, 2007). Victoria Falls, both the waterfall and the township, is the most popular tourist destination in Zimbabwe (Zimbabwe Tourism Authority, 2019). This is reflected in the number of foreign tourists who visited the two national parks located in the township of Victoria Falls (Zimbabwe Tourism Authority, 2019). These two national parks have the highest percentage of foreign tourists in Zimbabwe, 77% of the clientele to hotels in Victoria Falls were foreign, compared to a national average of 16% (Zimbabwe Tourism Authority, 2019). In Zimbabwe, there are differing patterns of tourism consumption for domestic and international tourists – with international tourists the key clientele for areas such as Victoria Falls and Hwange National Park which are both located in Matabeleland North.

Most tourists who visit Zimbabwe engage in some form of wildlife tourism; however, cultural tourism such as village visits have been suggested as a replacement for the 'big five' as a proposed more eco-friendly and sustainable form of tourism (Mutana & Zinyemba, 2013). There are, however, issues associated with this approach. The marketing discourses associated with mass tourism have created a tension in which tourism destinations must provide an 'authentic' experience, but one that can be replicated through processes of massification. Tourists often expect tourism experiences to represent the 'true' culture of the destination, and tourism often uses indigenous cultures and people to prove the 'authenticity' of the experience (Amoamo & Thompson, 2010; Burgold & Rolfes, 2013). Discourses of authenticity in mass tourism can further encourage the commodification of Indigenous cultures (Amoamo & Thompson, 2010; Enevoldsen, 2003). In Zimbabwe, Taru and Gukurume (2013) identify that commodification of culture is a double-edged sword; it helps maintain cultures whilst simultaneously ensuring that the culture cannot progress or change.

The commodification of culture is present in Matabeleland North. In terms of cultural representation in tourism, there have been concerns around the way food and culture are represented in online reviews of restaurants in Victoria Falls (Mkono, 2011, 2012, 2013a, 2013b). For example, food is 'othered' when tourists compare it to Minority World food or use adjectives to describe Majority World food, such as 'scary' or 'yuck' (Mkono, 2011). In online reviews on TripAdvisor, visitors to a cultural restaurant in Victoria Falls, Zimbabwe, depicted those who eat the local foods as 'daring' or not 'faint hearted' (Mkono, 2011). This is a form of othering and is a creation of a spectacle for tourists to observe. In the study by Mkono (2011, 2012, 2013b), foods were often marked as 'authentically African' which highlights the authority that Minority World tourists feel they have over images of Africa (to be discussed in later chapters). The language used to describe traditional food is a reproduction of the 'wild and savage' stereotype utilised by colonialisation but is also an example of the commodification of culture in tourism. Mass tourism has many issues associated with the manner in which it reproduces colonial dialogues and images.

## Summary

Zimbabwe has a turbulent economic history. The periods of hyperinflation and subsequent changes to currency have created serious mistrust in currency and a sense of economic uncertainty. Although still in a period of hyperinflation, Zimbabwe is recovering somewhat from the extreme period of hyperinflation in 2008–2009. This economic history has created a tumultuous present, where children may live with other relatives while their parents live outside of Zimbabwe. Parents struggle to pay school fees and there is a space for philanthropic foundations to intervene in schooling.

The resourcing of schools in periods of hyperinflation is fraught. The Zimbabwean education system has struggled to resource schools, in part due to the economic troubles but also due to the large increase in student enrolments following independence. Government schools in rural areas are under-resourced and government schools in all areas struggle to pay staff members. As a remedy, the introduction of satellite unregistered schools is intended to enable the development of schools by local communities. In practice, this has meant a range of funding sources for schools such as mines and tourism companies.

Zimbabwe's main drawcard and tourism industry is based around wildlife tourism. Cultural tourism has been suggested as an alternative to the 'big five', however, there are issues associated with this approach – which includes the proliferation of harmful discourses and a particular image of 'Africa'. With tourism ventures in Zimbabwe increasingly focussed on cultural sites, such as village tours, there are tourism providers who use charitable partnerships as leverage for entry to schools. These charitable partnerships are part of broader moves within tourism to 'give back' to communities and reflect broader trends of sustainable and community-based tourism.

In this chapter I have outlined economic and educational challenges, both of which create conditions of possibility for developmentourism to be considered as a viable funding option. As the Zimbabwean economy has experienced significant challenges, the government is unable to consistently fund education. The Zimbabwean education system has faced problems with staffing of qualified teachers, provisioning of textbooks and school buildings. These two challenges together have created a space in which international donors, through tourism or other avenues, have had the opportunity to intervene in the funding of schooling in Zimbabwe, and Africa more broadly.

**Note**

1 Although RTGS$ was not introduced until February 2019, reports from this year are in RTGS$.

**References**

Abraham, R. (2003). The localization of 'O' level art examinations in Zimbabwe. *Studies in Art Education, 45*(1), 73–87.

Amoamo, M., & Thompson, A. (2010). re(Imaging) Māori touism: Representation and cultural hybridity in postcolonial New Zealand. *Tourist Studies, 10*(1), 35–55.

Balint, P. J., & Mashinya, J. (2006). The decline of a model community-based conservation project: Governance, capacity, and devolution in Mahenye, Zimbabwe. *Geoforum, 37*(5), 805–815. https://doi.org/10.1016/j.geoforum.2005.01.011

Burgold, J., & Rolfes, M. (2013). Of voyeuristic safari tours and responsible tourism with educational value: Observing moral communication in slum and township tourism in Cape Town and Mumbai. *DIE ERDE - Journal of the Geographical Society of Berlin, 144*(2), 161–174. https://doi.org/10.12854/erde-144-12

Campbell, C., Andersen, L., Mutsikiwa, A., Madanhire, C., Skovdal, M., Nyamukapa, C., & Gregson, S. (2014). Children's representations of school support for HIV-affected peers in rural Zimbabwe. *BMC Public Health, 14*, 402. https://doi.org/10.1186/1471-2458-14-402

Chanakira, M., & Masunda, M. (2019). Corporate social responsibility engagement by Zimbabwean small and medium enterprises. *African Journal of Business Ethics, 13*(1), 62–77. https://doi.org/10.15249/13-1-217

Chilufya, A., Hughes, E., & Scheyvens, R. (2019). Tourists and community development: Corporate social responsibility or tourist social responsibility? *Journal of Sustainable Tourism, 27*(10), 1513–1529.

Chitiyo, G., & Chitiyo, M. (2009). The impact of the HIV/AIDS and economic crises on orphans and other vulnerable children in Zimbabwe. *Childhood Education, 85*(6), 347–351.

Chitiyo, G., Chitiyo, M., Rumano, M., Ametepee, L. K., & Chitiyo, J. (2010). Zimbabwe education system: Emerging challenges and the implications for policy and research. *Journal of Global Intelligence and Policy, 3*(3), 35–42.

Chitiyo, M. (2014). Challenges affecting the education of children in Zimbabwe. *Childhood Education, 90*(6), 414–417. https://doi.org/10.1080/00090456.2014.982971

Chung, F. (2014). Foreword: Integrating Zimbabwe's economic development. In E. Mamukwa, R. Lessem, & A. Schieffer (Eds.), *Integral green Zimbabwe: An African phoenix rising* (pp. xxvii–xxx). Gower Publishing Limited.

Coomer, J., & Gstraunthaler, T. (2011). The hyperinflation in Zimbabwe. *The Quarterly Journal of Australian Economics, 14*(3), 311–346.

Crush, J., & Tawodzera, G. (2013). The Perilous Trek: Zimbabwean Migrant Children and Teachers in South Africa. In L. Bartlett & A. Ghaffar-Kucher (Eds.), *Refugees, Immigrants, and Education in the Global South: Lives in Motion* (pp. 54–74). New York: Routledge.

Dekeza, C. (2018). *The responsiveness of rural primary schools to the needs of orphaned and vulnerable children in Zimbabwe* University of South Africa]. http://uir.unisa.ac.za/bitstream/handle/10500/25059/thesis_dekeza_c.pdf?sequence=1&isAllowed=y

Dekker, M. (2017). From macro to micro: How smallholder farmers in Zimbabwe are coping with dollarization. In W. A. van Beek, J. M. C. Damen, & D. W. J. Foeken (Eds.), *The face of Africa: Essays in honour of Ton Dietz*. African Studies Centre Leiden.

Dorsey, B. J. (1989). Educational development and reform in Zimbabwe. *Comparative Education Review, 33*(1), 4–58.

Dzimiri, P., & Runhare, T. (2012). The human security implications of Operation Restore Order on urban habitation in Zimbabwe. *Journal of Human Ecology, 38*, 191–205.

Enevoldsen, K. (2003). See No Evil, Hear No Evil: An "Outsider's" Encounter with Cultural Tourism in South Africa. *Cultural Studies ↔ Critical Methodologies, 3*(4), 486–502. https://doi.org/10.1177/1532708603253578

Govere, W. D., Tsokota, T., Chikuta, O., Mukewembi, A., & Chinofunga, P. (2013). The use of the internet to attract tourists to Zimbabwe. An analysis of the Zimbabwe Tourism Authority website. *International Journal of Management & Business Studies, 3*(1), 132–146.

Hanke, S. H., & Krus, N. (2013). World Hyperinflations. In R. E. Parker & R. Whaples (Eds.), *The Handbook of Major Events in Economic History* (pp. 367–377). Routledge.

Hlupo, T., & Tsikira, T. (2012). Comparative analysis performance of satellite primary schools and their mother schools in Masvingo Province, Zimbabwe. *Journal of Emerging Trends in Educational Research and Policy Studies, 3*(5), 604–610. https://hdl.handle.net/10520/EJC127701

Hove, M., & Ndawana, E. (2019). Education provision in the midst of a crisis: The Zimbabwean experience after 1999. *Journal of Peace Education, 16*(2), 215–216. https://doi.org/10.1080/17400201.2019.1627663

Imam, P. A. (2022). De-dollarization in Zimbabwe: What lessons can be learned from other sub-Saharan countries? https://doi.org/10.1002/ijfe.2177

Johnson, D. F. (1990). The politics of literacy and schooling in Zimbabwe. *Review of African Political Economy, 48*, 99–106.

Jones, J. L. (2010). 'Nothing is straight in Zimbabwe': The rise of the kukiya-kiya economy 2000–2008. *Journal of Southern African Studies, 36*(2), 285–299. https://doi.org/10.1080/03057070.2010.485784

Kanyongo, G. Y. (2005). Zimbawe's public education system reforms: Successes and challenges. *International Education Journal, 6*(1), 65–74.

Kanyongo, G. Y., Certo, J., & Launcelot, B. I. (2006). Using regression analysis to establish the relationship between home environment and reading achievement: A case of Zimbabwe. *International Education Journal, 7*(5), 632–641.

Kawere, S. M., & Dibie, R. (2000). The impact of the Economic Structural Adjustment Programs (ESAPs) on women and children: Implications for social welfare in Zimbabwe. *The Journal of Sociology & Social Welfare, 27*, 79–107.

Lines, T. (1988). Investment sanctions and Zimbabwe: Breaking the rod. *Third World Quarterly, 10*(3), 1182–1216.

Madondo, F. (2020). Perceptions on curriculum implementation: A case for rural Zimbabwean Early Childhood Development teachers as agents of change. *Journal of Research in Childhood Education*, 1–18. https://doi.org/10.1080/02568543.2020.1731024

Madziyire, G. T. (2015). *Evaluating the impact of philanthropic activities in public high schools in Mutasa district, Zimbabwe: An educational management perspective* University of South Africa]. https://core.ac.uk/download/pdf/43177311.pdf

Magudu, S. (2020). Configuring the key social justice concerns in rural education in Zimbabwe. In A. Masinire & T. M. Ndofirepi (Eds.), *Rurality, social justice and education in Sub-Saharan Africa Volume 1: Theory and practice in schools* (pp. 63–86). Palgrave MacMillan.

Makaudze, G. (2013). It still makes sense!: Shona ngano (folktale) and the contemporary Zimbabwean socio-economic and cultural setup. *International Journal of Development and Sustainability, 2*(2), 521–529.

Mandevani, K. M. (2014). Kushanya Mumamisha: Integral community activation via community-based tourism. In E. Mamukwa, R. Lessem, & A. Schieffer (Eds.), *Integral green Zimbabwe: An African phoenix rising* (pp. 67–86). Gower Publishing Limited.

Mangena, T., & Ndlovu, S. (2013). Decolonization-oriented naming of selected 'satellite' schools in Zimbabwe. ICONN 2.

Mangwaya, E., Jeko, I., & Manyumwa, C. (2013). Availability of print curriculum materials and its consequences for the quality of education in schools located on newly resettled farm areas in Zimbabwe. *Asian Social Science, 9*(1), 249–256.

Manwa, H. A. (2003). Wildlife-based tourism, ecology and sustainability: A tug-of-war among competing interests in Zimbabwe. *The Journal of Tourism Studies, 14*(2), 45–54.

Manwa, H. A. (2007). Is Zimbabwe ready to venture into the cultural tourism market? *Development Southern Africa, 24*(3), 465–474. https://doi.org/10.1080/03768350 701445558

Masaka, D. (2013). Pitting market economy against planned economy: A false dichotomy in Zimbabwe (1980–2011). *Journal of Black Studies, 44*(3), 314–329.

Maswera, T., Dawson, R., & Edwards, J. (2008). E-commerce adoption of travel and tourism organisations in South Africa, Kenya, Zimbabwe and Uganda. *Telematics and Informatics, 25*(3), 187–200. https://doi.org/10.1016/j.tele.2006.11.001

McGregor, J., & Pasura, D. (2010). Diasporic repositioning and the politics of re-engagement: Developmentalisng Zimbabwe's diaspora? *The Round Table, 99*(411), 687–703.

Meisenhelder, T. (1994). The decline of socialism in Zimbabwe. *Social Justice, Winter 1994*, 83–101.

Ministry of Primary and Secondary Education. (2017). *Annual education statistics profile Matabeleland North Province 2017*. http://mopse.co.zw/sites/default/files/public/ downloads/2017_Matabeleland_North_Province_Statistics_Profile.pdf

Ministry of Primary and Secondary Education. (2019). *2018 Primary and secondary education statistics report*. Ministry of Primary and Secondary Education. Retrieved November 10 from http://mopse.co.zw/sites/default/files/public/downloads/2018%20 Annual%20Statistics%20Report.pdf

Ministry of Primary and Secondary Education. (2020). 2019 Primary and Secondary Education Statistics Report. https://mopse.co.zw/sites/default/files/public/ downloads/2019%20Annual%20Education%20Statistics%20Report%20pdf% 20for%20UPLOADING.pdf

Mkono, M. (2011). The Othering of food in touristic eatertainment: A netnography. *Tourist Studies, 11*(3), 253–270. https://doi.org/10.1177/1468797611431502

Mkono, M. (2012). A netnographic examination of constructive authenticity in Victoria Falls tourist (restaurant) experiences. *International Journal of Hospitality Management, 31*(2), 387–394. https://doi.org/10.1016/j.ijhm.2011.06.013

Mkono, M. (2013a). African and western tourists: Object authenticity quest? *Annals of Tourism Research, 41*, 195–214. https://doi.org/10.1016/j.annals.2013.01.002

Mkono, M. (2013b). Augmenting foodservice experiences through cultural eatertainment at tourist destinations. *Journal of Foodservice Business Research, 16*(1), 40–51. https://doi.org/10.1080/15378020.2013.761019

Mlambo, A. S. (2017). From an industrial powerhouse to a nation of vendors: Over two decades of economic decline and deindustrialization in Zimbabwe 1990–2015. *Journal of Developing Societies, 33*(1), 99–125. https://doi.org/10.1177/0169796X17694518

Moyo, S., & Tichaawa, T. M. (2017). Community involvement and participation in tourism and development: A Zimbabwe study. *African Journal of Hospitality, Tourism and Leisure, 6*(1), 1–15.

Mudimba, T., & Tichaawa, T. M. (2017). Voices of local communities regarding their involvement and roles in the tourism development process in Victoria Falls, Zimbabwe. *African Journal of Hospitality, Tourism and Leisure, 6*(4).

Mudzingwa, K., & Magudu, S. (2013). Idealism versus realism: Expectations and challenges of beginning teachers in three districts of Masvingo Province, Zimbabwe. *Journal of Studies in Social Sciences, 3*(1), 33–54.

Mungazi, D. A. (1985). Educational innovation in Zimbabwe: Possibilities and problems. *The Journal of Negro Education*, *54*(2), 196–212.

Murithi, T. K. (2011). The United Nations: Between paternalism and partnership. In K. Kondlo & C. Ejiogu (Eds.), *Africa in focus: Governance in the 21st Century* (pp. 340–351). HSRC Press.

Mutambasere, T. (2022). Connecting religious transnationalism and development: Charitable giving amongst Zimbabwean Catholics in London. *Third World Quarterly*, *43*(1), 259–277. https://doi.org/10.1080/01436597.2021.2005463

Mutana, S., & Zinyemba, A. Z. (2013). Rebranding the Zimbabwe tourism product: A case for innovate packaging. *International Journal of Advanced Research in Management and Social Sciences*, *2*(4), 95–105.

Mutanga, C. N., Vengesayi, S., Gandiwa, E., & Muboko, N. (2015). Community perceptions of wildlife conservation and tourism: A case study of communities adjacent to four protected areas in Zimbabwe. *Tropical Conservation Science*, *8*(2), 564–582.

Mutema, F. (2014). An examination of learning conditions in Zimbabwe's satellite schools: A case of Somabhula resettlement area - Midlands Province. *International Journal of Humanities and Social Science*, *4*(8(1)), 284–290.

Muzapu, R., & Sibanda, M. (2016). Tourism development strategies in Zimbabwe. *Management*, *6*(3), 55–63. https://doi.org/10.5923/j.mm.20160603.01

Ndlovu, J., & Heath, E. (2013). Re-branding of Zimbabwe to enhance sustainable tourism development: Panacea or villain. *African Journal of Business Management*, *7*(12), 947–955. https://doi.org/10.5897/AJBM12.1201

Nhavira, J. D. G. (2019). The socio-economic impact of corporate social responsibility on the Zimbabwe mining industry. In S. Mugova & P. R. Sachs (Eds.), *Opportunities and pitfalls of corporate social responsibility: The Marange diamond mines case study* (pp. 49–72). Springer Nature. https://doi.org/10.1007/978-3-030-17102-5_4

Novelli, M. (2016). *Tourism and Development in Sub-Saharan Africa: Current issues and local realities*. Routledge.

Noyes, A. H. (2020). *A new Zimbabwe? Assessing continuity and change after Mugabe*. RAND Corporation. https://apps.dtic.mil/sti/pdfs/AD1097644.pdf

Nyahunzvi, D. K. (2013). Come and make a real difference: Online marketing of the volunteering experience to Zimbabwe. *Tourism Management Perspectives*, *7*, 83–88. https://doi.org/10.1016/j.tmp.2013.04.004

Nyahunzvi, D. K., & Njerekai, C. (2013). Tour guiding in Zimbabwe: Key issues and challenges. *Tourism Management Perspectives*, *6*, 3–7. https://doi.org/10.1016/j.tmp.2012.11.003

Riddell, A. (1998). Book review: Policy, evaluation and leadership: The context of educational change in Zimbabwe. *Compare: A Journal of Comparative and International Education*, *28*(1), 115–122. https://doi.org/10.1080/0305792980280110

Sithole, F. (2017). *Strategies to enhance teacher motivation in satellite secondary schools in the resettlement areas of Matabeleland North province* Zimbabwe Open University]. https://www.academia.edu/67625209/Strategies_to_enhance_teacher_motivation_in_satellite_secondary_schools_in_the_resettlement_areas_of_Matabeleland_North_Province

Snyman, S., & Spenceley, A. (2019). *Private sector tourism in conservations areas in Africa*. CAB international.

Southall, R. (2018). Bob's out, the Croc is in: Continuity or change in Zimbabwe. *Africa Spectrum*, *52*(3), 81–94.

Tarisayi, K. S. (2019). A school in distress: The manifestations of poverty at a selected satellite school in Masvingo district, Zimbabwe. *Journal of Geography Education for Southern Africa*, *4*(2), 79–96.

Tarisayi, K. S., & Manik, S. (2019). The role of land reform beneficiaries and the reasons for them developing and supporting a satellite school in Masvingo, Zimbabwe: A social capital marriage of Nhimbe and allied reasons. *Education as Change, 23*(1), 1–18.

Taru, J., & Gukurume, S. (2013). Corporate responsibility in the tourism industry: Lessons from communities surrounding Great Zimbabwe monuments. *Journal of Tourism Challenges and Trends, 6*(1), 59–74.

Tawodzera, M. C., & Themane, M. (2019). Schooling experiences of children left behind in Zimbabwe by emigrating parents: Implications for inclusive education. *South African Journal of Education, 39*(Supplement 1), S1–S13. https://doi.org/10.15700/saje.v39ns1a1797

Tichaawa, T. M., & Mhlanga, O. (2015). Residents' perceptions towards the impact of tourism development: The case of Victoria Falls, Zimbabwe. *African Journal of Hospitality, Tourism and Leisure, 4*(1), 1–15.

Trading Economics. (2020). *Zimbabwe tourist arrivals.* Retrieved November 12 from https://tradingeconomics.com/zimbabwe/tourist-arrivals

Utete-Masango, S. J. (2016). *Education sector strategic plan 2016–2020.* Online: https://www.globalpartnership.org/node/document/download?file=document/file/2016-07-education-sector-strategic-plan.pdf

World Bank. (2020a). *Gini index (World Bank estimate).* Retrieved November 17 from https://data.worldbank.org/indicator/SI.POV.GINI

World Bank. (2020b). *School enrollment, primary (%gross) - Zimbabwe.* Retrieved November 10 from https://data.worldbank.org/indicator/SE.PRM.ENRR?locations=ZW&most_recent_value_desc=false

World Bank. (2020c). *School enrollment, secondary (% gross) - Zimbabwe.* Retrieved November 10 from https://data.worldbank.org/indicator/SE.SEC.ENRR?locations=ZW&most_recent_value_desc=false

Zapata, M. J., Hall, C. M., Lindo, P., & Vanderschaeghe, M. (2011). Can community-based tourism contribute to development and poverty alleviation? Lessons from Nicaragua. *Current Issues in Tourism, 14*(8), 725–749. https://doi.org/10.1080/13683500.2011.559200

Zimbabwe Tourism Authority. (2019). *Tourism trends and statistics report.* Tourism and strategic research division. http://www.zimbabwetourism.net/tourism-trends-statistics/

ZIMSTAT. (2014). *Facts and figures 2014.*

ZIMSTAT. (2017a). *Education report 2017.* ZIMSTAT. http://www.zimstat.co.zw/wp-content/uploads/publications/Social/Education/Education-Report-2017.pdf

ZIMSTAT. (2017b). *Zimbabwe poverty report 2017.* ZIMSTAT. http://www.zimstat.co.zw/wp-content/uploads/publications/Income/Finance/Poverty-Report-2017.pdf

ZIMSTAT. (2019). *Quarterly digest of statistics.* Retrieved from https://www.zimstat.co.zw/wp-content/uploads/publications/Miscellenious/Digest/2019/Digest_Q4_2019.pdf

ZIMSTAT. (2020). *Poverty and social impacts of COVID-19: Results from the rapid PICES phone survey data.* http://www.zimstat.co.zw/wp-content/uploads/publications/Income/Finance/RAPID_PICES_Wave1.pdf

# 3  Introducing Matopo School

I am sitting in the front seat of a safari vehicle in a national park outside of Bulawayo, Zimbabwe's second largest city. We are speeding along a dirt road in a modified Land Rover in search of rhinoceros that have been spotted in the area earlier. They are highly endangered and anti-poaching squads guard them. Our tour guide has text messaged the poaching squad to see if they will tell us where the rhinoceros are. When we arrive, I see him hand one guard some US bank notes – an example of Zimbabwe's informal economy at work – I wonder how many groups are brought through this area under the guise of 'finding' rhinoceros. Perhaps we are the tenth group today. I am on a brief tour around Zimbabwe, as part of a group package and so far, have seen the classic 'African' sites: Victoria Falls, animals on safari, and now I am walking with rhinoceros. After the encounter with the rhinoceros, we begin making our way back to the vehicle, and the guide asks why I am in Zimbabwe. I am usually asked this when I am alone doing a tourist activity – it seems most people find it unusual for a young woman to be travelling alone. I mention my research, and the tour guide immediately begins to tell me about a local school that his company sponsors and supports. Every person I talk to in Zimbabwe seems to have a story about sponsoring a school or working with a school. If they are not personally involved, they know someone who is.

In the area I work and research, other schools in the area also have tourists who visit to varying degrees. I hear reports of classroom blocks built by tourism operators, of local people building their own school, and of donations of uniforms. One teacher who worked at Matopo School previously had a child provided with a uniform from tourist donations. The teachers at Matopo School frequently tell me stories as a cautionary tale about another local school for whom tourism did not work. I also hear stories of volunteer tourists who have worked in these schools, or international school groups who have visited schools in the area. Quite near to Matopo School, there is another school who receives tour groups in the same manner as Matopo School; in fact, one of the companies who visit Matopo School visit several other schools in the area. It is unsurprising, then, to learn that 95% of Matabeleland North tourism operators interviewed by Mutana et al. (2013) had paid school fees of children and

DOI: 10.4324/9781003407980-3

several others had donated other pieces of infrastructure, such as school buildings. In Zimbabwe, philanthropy and schooling go hand in hand.

As detailed in the previous chapter, due to the ongoing economic problems,
many schools face challenges in providing facilities to students. Outside of
urban areas, schools are incredibly under-resourced. In the context of the
ongoing economic problems, the Zimbabwean education system is also in the
midst of a prolonged crisis (Chitiyo & Chitiyo, 2009; Chitiyo et al., 2010; Hove
& Ndawana, 2019), which has been exacerbated by COVID-19 shutdowns
(Dube et al., 2021; ZIMSTAT, 2020). These problems included a lack of textbooks, trained teachers, and infrastructure (Madziyire, 2015). Teachers in government schools are underpaid. As a remedy to the resourcing problems, many
schools use philanthropic donations as a source of funding. For some schools,
these wealthy donors are sourced through tourism ventures.

The failing economy causes issues beyond resourcing of schools. It is within
this context that student sponsorships remain valuable as families take financial and moral responsibility for children who may not be their own, such as
their grandchildren, nephews, and nieces (Tawodzera & Themane, 2019). The
high orphan rate, coupled with high unemployment, makes school fees inac-
cessible for many Zimbabwean families but particularly those who are caring
for orphaned children. Although some parents leave Zimbabwe in hope for a
better future and send remittances back to their families, not all do. For example, in the community who participated in this study, four children were in the
care of their grandmother after their mother had crossed the river to sell items
in Zambia and did not return. The grandmother was responsible for the children's welfare and relied on sponsorship of the children to pay school fees.
They were not the only children living under guardianship.

With Zimbabwe's economic decline, there is very little government support
for the education system. As part of the growing desire to 'giveback', in
Zimbabwe the linking of non-government organisations (NGOs) and tourism
is common (Mutana & Zinyemba, 2013; Mutanga et al., 2015). Projects which
combine tourism and philanthropy provide infrastructure and support for the
school system – which previously has not been provided through government
funding. The issue of school tours is becoming more pertinent; as tourism-
orientated NGOs are becoming more prevalent in countries in the sub-Saharan
Africa (Clausen, 2019; Wondirad et al., 2020), visiting schools as part of a
philanthropic tourism exercise is also becoming more frequent (Chilufya et al.,
2019). Scarth and Novelli (2019) identify visiting a school as part of a mass
tourism package as a type of travel philanthropy that results from the 'compassionate by-products of a wildlife safari' (p. 90).

## Matopo School

Matopo School is a primary school established in 2013. Matopo School is, and
always has been, a site for tourism. The school was established by the founders
of a tourism company, with a view to 'giveback' to the local community and

support the education of children in the area. Shumba Safaris is the company run by the school founders, Alison and Thompson, and the guests on these tours book bespoke tours which usually include a visit to Matopo School. In some form, the school has always been funded by tourism. It is located on a relatively well-maintained road and is only a few kilometres from a tarred road which connects to a regional centre. The teachers and students consider it a rural school. The school is close to the nearest regional centre, which is a smaller town with a population of roughly 35,000. In 2019, Matopo School had nine teachers and four support staff. Since 2013, they have added one grade each calendar year and in 2019 the school ranged from Early Childhood Development to Grade 5.

Matopo School is located in Matabeleland North, Zimbabwe. To provide insight into the condition of schools in Matabeleland North, I now present some statistics. This is not to essentialise or mark the region as developed or under-developed, but rather to situate the extreme benefits that Matopo School receives from tourist dollars. Across Zimbabwe, in 2018, 45% of schools did not have access to any electricity source with Matabeleland North province having the highest number without electricity (Ministry of Primary and Secondary Education, 2019). In Matabeleland North in 2017, 82% of primary schools did not have access to any source of electricity (Ministry of Primary and Secondary Education, 2017). Access to electricity highlights the rural/urban divide in which schools in urban areas are more likely to have electricity (Ministry of Primary and Secondary Education, 2019). In Matabeleland North in 2018, 11% of schools had access to water through piped water sources and 66% had access to water through boreholes (Ministry of Primary and Secondary Education, 2019). Across Zimbabwe, 20% of schools had internet connectivity in 2018, with Matabeleland North province the least connected, at 10% (Ministry of Primary and Secondary Education, 2019). Internet connectivity represents another rural/urban divide, with Bulawayo (63%) and Harare (86%) having much higher levels of connectivity (Ministry of Primary and Secondary Education, 2019). Matopo School has running water, electricity, internet connectivity, a computer lab, and flushing toilets, all obtained through donor funding. Despite having electricity, it can be unreliable and there are frequent power outages. At the time of my research, the internet was connected but did not work most of the time. As of 2023, there is also a medical clinic and secondary school located on the school grounds.

The school is a satellite school in the process of registration with the government and has facilities such as a photocopier and a computer lab. As Matopo School is a non-government school, there is a flexibility in the administration of funds and in the spending of charitable donations; however, they also face the challenge of needing to source funding – rather than relying on government funding. In contrast to other schools in the rural area, the school has many facilities, including a consistent feeding programme for all children at school, regardless of payment of school fees. At the time of the fieldwork, Term 1, 2019, the school had 221 children enrolled with fairly equal enrolments of boys and girls.

Each tour has a similar configuration. The tour begins when the tourists arrive on the mini bus. When they arrive, selected children greet them at the minibus and take them by the hand to a nearby area of the playground. Then a group of children sing and dance for the tourists, which includes a mix of Ndebele songs/dances and some songs in English. If the whole school is present, each of the teachers are introduced at this time. A school history is then provided by either the principal, school founder, or the teacher assigned for duty. During the school history there is a direct appeal for sponsorship of children and donations. Following the school history, each classroom and part of the school is visited by the tour group who are either guided by children or by the teacher on duty. They will then either visit a nearby village or leave the school.

The school received 126 visits in the 2019 school year, from three companies: Education Tours, Adventure Company, Shumba Safaris. On average, there are roughly three visits a week, but the high season (May–October) has more per week, whilst the low season (November–April) has less. All companies visit the school as part of a larger tour of Southern Africa. For example, Education Tours and Adventure Company offered several tours with different configurations and not all tours attended Matopo School, with other schools in the region also receiving tourists from these companies. The tours for Education Tours and Adventure Company were all-inclusive small group tour packages, which cater to American tourists over 50. During the period of fieldwork, all visitors to the school were American, and appeared to be white. Although black African American tourists were uncommon, Tabitha, a teacher at the school, identified, 'we have one or two black Americans still coming here'. The demographic makeup of the tourists, therefore, was predominantly white with very little cultural variation.

The three companies are different in their aims and target demographic, so a detailed explanation is provided here to help clarify how each tour runs. Figure 3.1 shows the process for each tour visiting the school.

The components of the tour were also depicted in a drawing by Aysha (Figure 3.2).

Aysha's drawing shows the school building, with each room including a welcome greeting ('Welcome to the office', 'welcome to Grade 4 class' etc.); the singers, dancers, drummer, and the visitors watching the performance; and the tour bus.

Given the multiple companies and people involved in the implementation of philanthropic funds at Matopo School, I now detail the various companies, their funding contribution, and the main people involved in the funding.

**Education tours**

Education Tours visited Matopo School 41 times in the 2019 calendar year. Their visits were time consuming for the school as the teacher on duty would also take the tour group to a local village, leaving their class unattended for the

| Education tour | Adventure Company | Shumba Safaris | Backpacker |
|---|---|---|---|
| Pre-determined arrival at school. | Pre-determined arrival at school. | Pre-determined arrival at school. | Arrive unannounced. |
| Tour bus parks and children take each tourist by the hand and lead to entertainment area. | Tour bus parks and children take each tourist by the hand and lead to entertainment area. | School founder arrives with tourists. | Visit the office |
| Whole school greets the tour group, performs songs and dances. | 'Entertainment group' greets tour group, performs songs and dances. | School founder introduces tourists to whole school. School performs songs and dances. | Donations received with quick tour undertaken, no performances by school. |
| Tour group moves to the shed for school history by teacher on duty and hand over of donations. School returns to classrooms. | Tour group moves to the shed for school history by teacher on duty and hand over of small donations. Student tour guides remain behind to wait. | Tour group moves to the shed for school history by school founder and hand over of small donations. School returns to classrooms. | Tourists leave. |
| School tour conducted by teacher on duty. Tourists enter classrooms and may tutor individual children or teach a portion of a lesson. | School tour is conducted. Each classroom is visited with student guides supervised by teacher on duty. | School tour is conducted by school founder, with long visits in each class. | |
| Tour group leave for village tour, guided by teacher on duty. They visit a homestead and taste traditional food. | Tour group leave school for a village tour conducted separate from the school community. | Tour group leave. | |

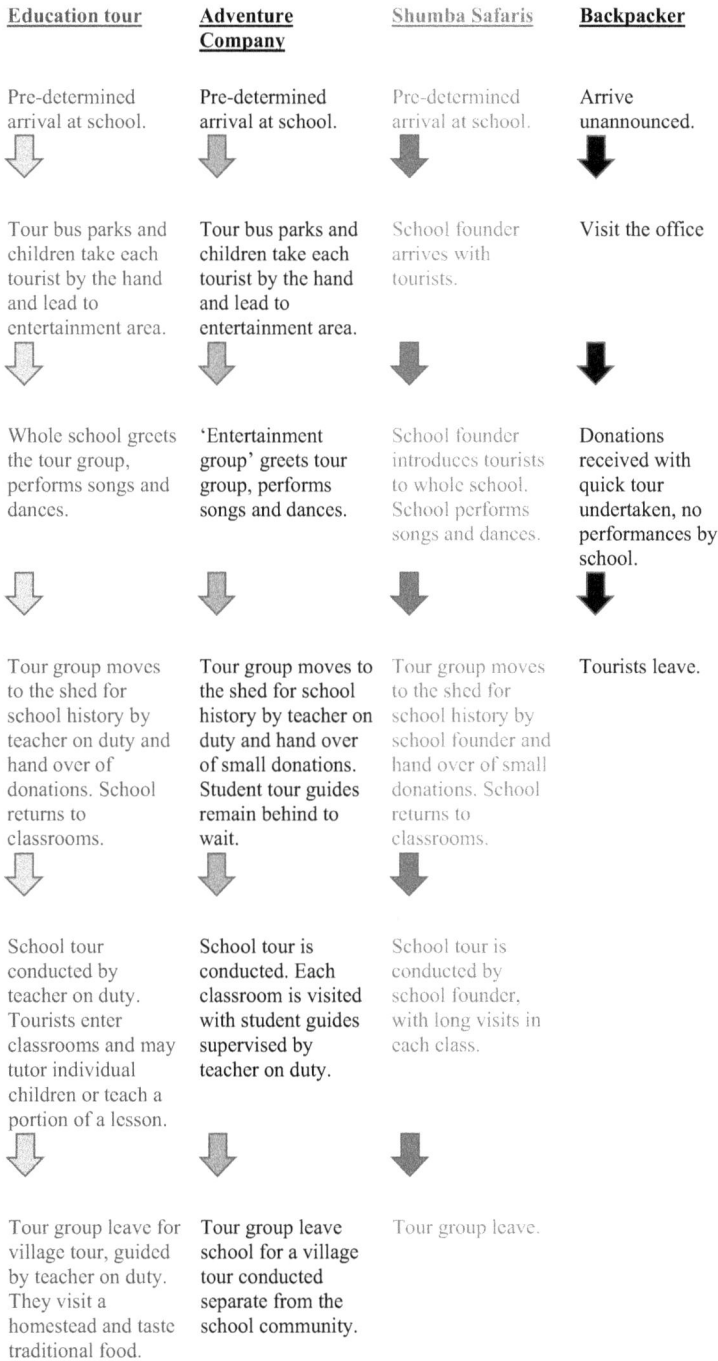

*Figure 3.1* School tour diagram.

*Figure 3.2* Drawing by Aysha depicting school tour.

period of time they were not on school grounds. Education Tours describes itself as offering 5,000 tours across 150 countries and as a not-for profit educational tour company. Their focus is combining education and tourism, with a suggested reading list provided to tour participants prior to departure. Greg, a tour guide for the company, who lived in the local area explained this to me,

> You know there are many companies that you can come to Africa on tour – but the focus that Education Tours has is education. So, it is adult learning. When I do tours of South Africa you get a history of Cape Town, a background to Apartheid, you go to the Apartheid museum, you have a local step on guide who talks everybody through it. I have got to give a series of a lectures throughout the journey, and it could be on lions, termites, Zimbabwe it could be anything, elephant, social life of the elephants, conflict with elephants, that sort of thing. So, the whole focus is on education. And one of the things is that they try and do is give back to the community.

From my observation, most of the customers of Education Tours were American and retired. Greg confirmed this saying,

> So, when we go out to the school it is part of the give back. The majority of our visitors are American, very generous.

On each visit the Education Tours group would donate $200/US cash directly to the school, which is used for larger purchases and emergencies, which meant there was very little administration involved and very little oversight on the part of the donors. The tour guide would give the cash to the school principal and sign a register book to acknowledge the receipt of the donation. As an example, the school had used these funds to pay for emergency treatment of a student who had fallen ill and also to purchase a large freezer for the school feeding programme. The tour group schedule was provided to the school founders, Thompson and Alison, who would provide it to the school. The school founders seemed to have a level of control over the number and design of the visits.

### Adventure Company

Adventure Company describes itself as a small-group tour company which focuses on places 'less travelled'. Adventure Company offers tours in over 80 countries and emphasises their local connections and the value they provide to their customers. Adventure Company explicitly identify on their website that their target demographic is Americans who are over 50 years old and physically active. Matopo School received 80 visits from this company in the 2019 calendar year. The visits from Adventure Company were less time consuming for the school as only specific classes were required to perform; however, each tourist was guided by a child from the school. The village tour was conducted by the Adventure Company tour guide. Adventure Company has a charitable organisation associated with the tour company and donations from this company were organised through a local NGO employee called Mary. Alison, a founder of the school, identified that the largest benefit from these tours was an increased interest in child sponsorship, but they had not yet received substantial donations as a result of the tours.

The tourism employee and a former school principal, Mary, assists the Adventure Company to implement the funding of their charitable organisation in Matopo School and other local schools. At the time of this research, Mary had recently begun work with Matopo School, commencing in 2018 with a donation of books and pencils; but she had been working in other local schools since 2009. Mary describes herself as wearing two hats, the first being her work with an NGO associated with Adventure Company and her work with Luxury Tours, who also have a charitable organisation. Luxury Tours did not visit the school, although they pay Mary's salary.

Mary's current role at the Luxury Tours charitable organisation began as an implementer of funding for educational camps for children, funded by Luxury Tours. She describes the transition from NGO work to a more sustained programme of work in schools,

> We thought we needed to reach more people through educational clubs rather than just the few we influenced through camps. We thought it would be better if we started doing educational clubs in schools. So that is when my eyes really opened because I then went into schools and this

was about 2009. I suddenly realised, that **[pause]** it was almost... it was limping along, the education system in Zimbabwe. And there were a number of factors on this. The economic demise of Zimbabwe and we had very hungry children. You went to the schools and they were actually washed out, hungry, were not learning. They would walk off to school each day and all that, so that is when I thought well you know they cannot learn. Let us get nutrition programmes going. And then it went onto well, okay we need to de-worm them, because they have all got worms... which is a very common problem in Zimbabwe – well in rural areas.

Then, it is, oh, well the whole system is not allowing us to run educational clubs because of that. So, we need food, then we looked at the children – then they need water, close on hand, then they need sanitation, and all that. We started putting in boreholes and what have you. Sort of getting the whole environment right for that. And then again, we had a huge dip, and we had a mass exodus of teachers, leaving the schools. Suddenly we had made this whole situation where everyone was coming to school and, 'yay' and you know, got a meal and what have you. Now suddenly all the teachers are gone. Now what are we going to do? So that is when we started looking at subsidising teachers and getting in school leavers to take positions. And, then we realised that they needed some training. We started facilitating that and the sheer thing of huge classes, so few teachers, it is heartbreaking.

Mary describes the complex problems associated with schooling in Zimbabwe, identifying the multiple problems she had to solve prior to beginning to implement educational clubs. Her role at Adventure Company came as a part of her work in schools in the region.

Mary is a white Zimbabwean woman. She was a teacher and school principal prior to moving into local NGO work; as a teacher she worked in government and private schools. When describing her own experiences in local schools she positions it as difficult, referencing government policies and the difficulty of teaching in tough economic situations. She chose to move from teaching as she described it, 'I have always found that teachers end up taking the can when an economic situation drops'. By 'taking the can' she means teachers face problems when the economy falters, such as lowered incomes with expectations remaining the same, lack of resources, and large class sizes. Her choice to move out of teaching reflects her relative privilege in Zimbabwe at a time that many were forced to leave the country to find work. Of note, she left teaching prior to the 2008 economic meltdown in which most schools closed for a year. Her work in schools in the present is positioned through her own experience of working in both rural and private schools. Throughout her interview she asserted her authority several times by stressing that her knowledge is based on her prior experience in schools and the hardships she encountered in her own teaching experience.

**Shumba Safaris**

Shumba Safaris caters to high-end tourists who could afford to spend large amounts of money on safari and who seek luxury tours. Shumba Safaris was founded by Alison and Thompson. They have no formal teaching experience. Alison is based in the United Kingdom and met Thompson on a safari tour in the early 2000s. Alison is a white woman who has previous experience in corporate accounting. Thompson is a black Zimbabwean tour guide from the Masvingo Province who has previously spent time working as a guide in both Matabeleland North and areas of Botswana.

Prior to opening Matopo School, Thompson had purchased land nearby and witnessed children walking long distances to school, or not going to school due to the distance, so he decided to create a play area to help support parents and children. The idea morphed into an Early Childhood centre, which then became a school. Thompson details the process of how the school began:

I had always wanted to build some little home, maybe something for orphans or something. There was something in me from the way I was brought up. My parents were very soft people, they ended up living with all strangers from everywhere, but part of me felt like I had been very lucky. Considering most people of my generation are probably… most of them are dead. And, most of them were cut off earlier because schools shut down during the war, and that was the end of it. So, I felt being so lucky, I should also maybe try and look after other people and that is always been at the back of mind. How I was going to do it, I had no idea. I knew I needed to do something, so first I wanted to build a little orphanage. Then I came into the village, I wanted a place that I could call home, which is like back home. And they gave me a place, and I look at all the feasibility, what we could do, you know. Somebody told me, somebody had just started an orphanage nearby. I thought, okay let's throw that idea away, the next thing is going to have to be a school.

I wanted a little play centre. It wasn't meant to be a preschool, it was a play centre to get the kids away from, you know all the abuse at home. You know the kids being asked to do manual work and stuff and they were running around in the village doing nothing. I thought, well if we have a play centre, maybe we can steal them away from the parents and keep them busy. But then the play centre turned into a place where I had one teacher, she was the one there, overseeing the kids play. Then she started doing a couple of lessons, then just went from there to what it is now. Yeah. Yeah. Thats how it all happened.

Again, like I say, always say that it's just the dream that went wrong. You know, you think 'maybe let me try', it wasn't intended to… Cos I knew there was a chance I might not get money, funding. I sat with Alison, looked at how we would do the whole thing. It was like, okay maybe we have been successful on the safari side, but why don't we take the opportunity to utilise the guests that we meet? So that they can help us with the project here and get something.

And one thing led to another, and then Alison came up with, 'well, why don't we register a charity?' And I was like, 'oh god, now we looking for the most difficult things to do'. And it worked out, she said 'registration is done'. We became the first two directors, then we realised a third person so we got Alison's husband involved as well. You know the American market, which is 99% of our business, turned out to be so positive about the idea of the charity and it has been from people we know, who have gone and found their relatives and friends. And now it has become a great success. We do also make sure that none of the money donated is abused by using in house management, that's why I work all the time. Because we don't employ anybody, I don't get paid for doing charity work, because I have a full time job. So 100% of our contributions received are used directly towards the project. The company, Shumba Safaris, will subsidise any that needs extra. So we get a little extra from our business, and we plough that to help the charity as well.

Through Shumba Safaris, Alison and Thompson run a registered charitable organisation in the United Kingdom, with the sole purpose of funding Matopo School. The charitable organisation created through Shumba Safaris now pays seven teacher salaries and bolsters the income of the two government teachers allocated to the school.

Alison, who organised the tours, would usually include a school tour in every itinerary. The school received five visits from this company in the 2019 calendar year. I did not observe a visit, as none were conducted in Term 1 of school. This is probably due to the fact that the high season for safari tours is July to September. I did, however, witness a staff meeting which explained that in August one of the visits from a Shumba Safaris group would fall on a Sunday. It was requested that all children and staff attend on this day and act as if school were in session, like a normal day. The people who were visiting on this tour paid for the borehole to be constructed and currently sponsor five children at the school.

**The funding arrangement**

Matopo School has three sources of funding: Education Tours who provide $200/US cash per visit and larger donations; Adventure Company who donate through a charitable organisation associated with their company; and Shumba Safaris which is a travel agency/tour company owned and operated by Thompson and Alison (the linking of NGOs with tourism ventures is common – see Mostafanezhad, 2014). Figure 3.3 illustrates the relationships of funding, for ease of understanding.

Each tour group would also bring small items such as pens and pencils as all three companies promoted the practice of 'pack for a purpose'. 'Pack for a purpose' is a movement which recommends packing small items into luggage

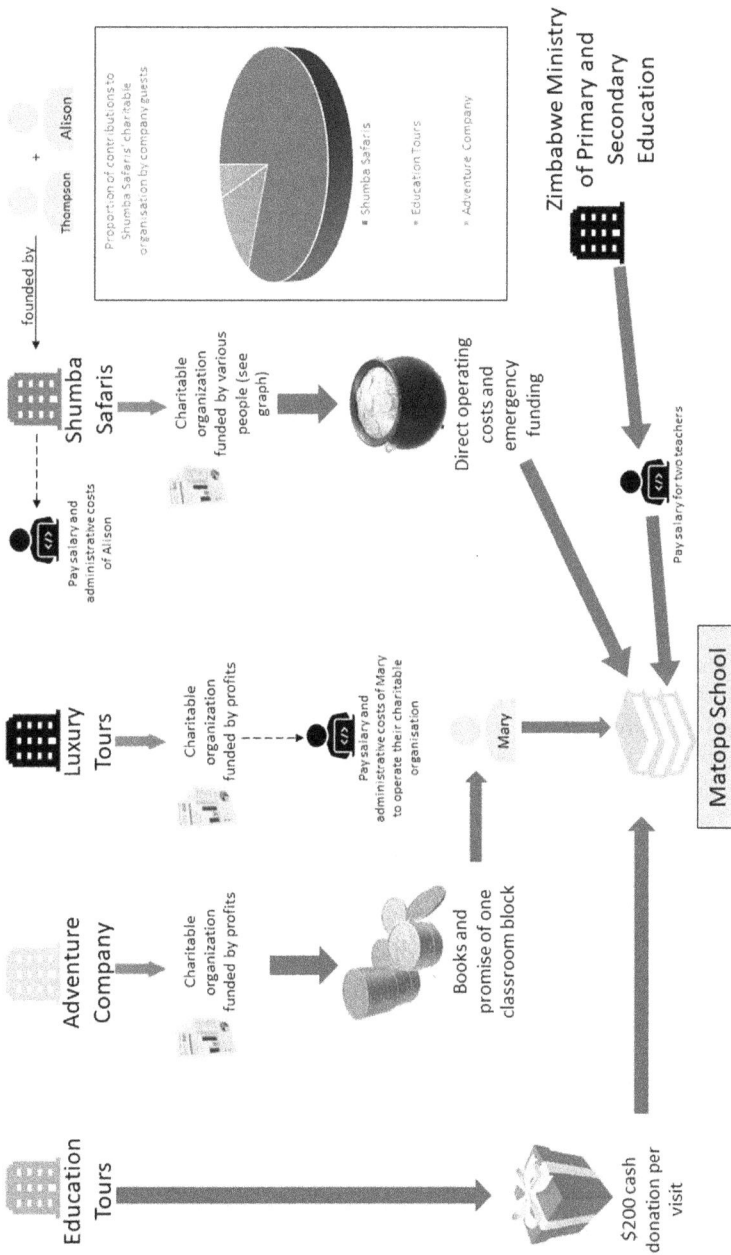

*Figure 3.3* Funding arrangement at Matopo School.

to give to local people in your final destination. Gift-giving as part of tourism is under-researched, and the extent to which it occurs is unknown (Scarth & Novelli, 2019). A school founder, Alison, describes the funding situation,

> Most of it at the start, and then going forwards, has been from either personal donations or a handful of our guests. A few have been early Education Tours guests who have remained very very supportive. But most is actually Shumba Safaris guests and very little is Adventure Company guests. Very little at all. See 71% of funding to date, has been personal and guests of Shumba Safaris. 12% has been from Education Tours[1], 4.5% has been Adventure Company. And the Adventure Company has mainly come from one individual who visited last year and came back and wanted to do a few projects, like the toilet block.[2]

As Alison describes, the school is funded from a variety of sources, who provide different levels of funding. The funding arrangement at Matopo School was far from straightforward, with multiple players and companies involved. The multitude of arrangements appears to be relatively common, from anecdotal observations of other schools in the area, who also rely on multiple forms of tourism partnerships for funding or support.

**My relationship to the school**

There is no denying there is a level of irony involved in writing a book critiquing neocolonial practices when I am the grandchild of former colonists and working from a position of relative privilege in Australia. I attempted at every stage of this project to engage with and consider the situated context in which I was working. I reflected deeply on my own whiteness and complicity in hegemonic capitalist structures. Throughout the research project I kept a file with my reflections of my own position and the work I was encountering. As an example, I have provided an early extract here:

> *I have progressed through my postgraduate studies and encountered the works of Spivak (1998) who in particular, announces that the subaltern cannot speak. As such my own whiteness and privilege were thrown into the spotlight. I have worked to navigate the complexities of this situation by engaging with the work of Chilisa (2012), who details the ways in which whiteness can work in postcolonial spaces. Yet, I still feel a strong urging of 'do not do this'. Reading the work of Fine (1994) it becomes abundantly clear that this space is not easy to navigate, 'if poststructuralism has taught us anything, it is to beware the frozen identities and the presumption that the hyphen is real, to suspect the binary, to worry the clear distinctions' (p. 80). So how do we move forward from here?*

While this extract is insufficient in exploring the ethical considerations needed for this study, it should highlight the complexities of researching in a space that is not your own and the thinking work I undertook behind the scenes. This 'thinking work' is not always reflected in institutional approvals.

A key dimension of ethnography is a flexible approach to research planning and implementation (Hammersley, 2017). A flexible approach means being open to changes in research design, and working in conjunction with communities you are researching in. As part of the role of reciprocal relationships (Chilisa, 2012), I followed the school's guidance on how they wanted me to act in the school. The school treated me as another teacher, and I was expected to follow the norms of behaviour for a teacher at a Zimbabwean school, such as dress code, arrival times, and behaviour, rather than my initially intended role of an observer/least adult (see Harper & Smithers, 2023).

Despite being considered one of the teaching staff, in many ways I acted as a teaching assistant who sometimes filled in for the teacher. Abebe (2009), working in the Ethiopian context, refers to this role as the 'atypical adult teacher' (p. 459). As such, I was not quite 'least-adult', but I was not quite a typical teacher. In my role as 'atypical adult teacher' at the school, I participated in school life by lining up with teachers during assembly, taking class attendance in the mornings and marking student work for the teacher I was paired with. On days when this teacher was absent, I was expected to take over the teaching of the class (see Harper & Smithers, 2023). The teacher I was paired with, Zibusiso, engaged me in collaborative discussions about events that occurred in the classroom and the ways that we could improve student understanding of topics.

In my 'atypical teacher' role I participated in various parts of the school life. Each day at school I followed the lead of others around me, regarding where I should be for the day, or what I should be doing. For instance, at one time the school needed to gain authority to cut a tree to build a traditional display hut for heritage studies. I was instructed to join Zibusiso and the school principal on the trip to gain permission. In this instance, the school was actively leveraging of my position as a white, Minority World woman. During the car ride to the office, I was informed they thought they were more likely to get permission if I was with them. In perhaps a contradictory turn of events, and a reflection of my lack of community ties, the school principal's husband who was a church leader in the area was a more useful connection. In my role as 'atypical teacher' I also performed the role of a secretary, typing reports and letters for the principal. In another instance, I helped cook and serve food for other school teachers who came to the school for an area meeting. As I have demonstrated through these examples, my role was ever shifting in the school and based in an understanding of reciprocity; I aimed to be as useful to the school as possible.

During the school day, if I had the opportunity, I wrote field notes in a notebook. Children and teachers became used to me scribbling these notes. Some teachers actively participated in my notetaking; they would show me things so that I could write them down for reference later. For example, in the second month there was a funeral held in the nearby village. The teachers insisted I go, so I could write about the funeral. At this funeral I participated in the ceremony and learnt how funerals work in the village. Although not very relevant to my research, I am now writing about it here to reflect the importance of this moment for me, and the value the research participants felt it had. At other times, I was shown WhatsApp threads about the current state of politics or given detailed explanations about certain parts of the school, with the expectation that I would be detailing it in my notebook.

These field notes were often very rough and intended to be prompts for me to explore in-depth later, as at times it was hard to write detailed notes (this is an established technique see Walford, 2009, p. 120). For example, the teacher was often out of the classroom, and there was an expectation that I would teach or mark books at this time. During these periods, writing in-depth notes was almost impossible. As part of the messiness of fieldwork (Ellingson, 2017), I accepted the notes as disjointed, and in the afternoon, after school, I would use the prompts to write more detailed notes. As the researcher cannot fully separate themselves from the process of writing field notes (Ellingson, 2017; Fine, 1993), at times the field notes became like a diary in which I could de-stress and figure out how I felt about the situations I found myself in. In particular, I focused on what it felt to be a female researcher alone in a foreign country and the specific methodological problems I faced.

As noted above, my field notes often focussed on the feelings I had about being a female researcher alone in a foreign country. There are well-established challenges in conducting fieldwork as a woman who is alone in a country (Brandt & Josefsson, 2017; Hanson & Richards, 2017; Pollard, 2009). Alongside the negative emotions that ethnographers usually encounter, such as fear, frustration, and disappointment, female researchers often face sexual harassment (Pollard, 2009). There were times during my fieldwork where I was approached in the street, or in restaurants, by men who would ask for my telephone number or where I lived. Sometimes, these men would follow me until I got in a taxi. On one occasion, I was on my way to Zambia with another (female) teacher and our regular taxi driver saw us walking on the street and yelled sexual comments, intended for me. In a similar manner to Brandt and Josefsson (2017), I faced questions such as, 'if you have a husband, why isn't he here with you?'.

Although not directly relevant to my experiences at school, these repeated encounters altered my emotions, and my field notes often reflected the changing emotions I was feeling such as, fear, discomfort, and insecurity. Further, due to these encounters on the street I was cautious in my interactions with the men in this study, as I was acutely aware of being a woman who lived alone; something that the participants in this study might not know but could assume. These feelings of apprehension about forming bonds with these research participants may have influenced the interviews I conducted with them. Brandt and Josefsson (2017) describe this sort of shift in behaviour as 'part of being in the field and adjusting gender and racial norms' (p. 29). In this way, I was also adjusting my sense of normal, as I navigated gender norms that were different to those in my home country, something that was reflected often in the way I wrote about my encounters.

Throughout my research I moved through varying positions as an insider or outsider. At times, the research participants identified me as an insider, through my qualifications as a teacher; however, I was also an outsider as I held a university qualification and was mostly experienced in Australian high school classrooms. In another example, one participant identified that, as I was born in Zimbabwe, I was Zimbabwean like her. This insider position, as declared by one teacher, is incredibly complex to navigate. In one sense, I *was* born in Zimbabwe, but I was born into privilege and spent most of my life in Australia. Despite the commonality of being born in the same country, our life histories have looked remarkably different. All of my relationships during the fieldwork were shadowed by race, and the pre-existing race relations that existed in Zimbabwe, even when I interviewed people who were also white.

Another complexity I navigated during my fieldwork was my own privilege. Despite reading extensively about privilege, whiteness, and class (for example Ahmed, 2007; Bonnett, 1997; Cabrera, 2016; Ellsworth, 1989; Faria & Mollett, 2014; Moreton-Robinson, 2002), prior to the fieldwork period. I was not prepared for the complexities of navigating being a PhD student conducting overseas fieldwork in a low-income country. At times the participants made appeals to me, regarding their own incomes or sponsorship status of their children. These appeals usually included me funding these things or finding someone in Australia to fund then. My white skin marked me as wealthy (Lozanski, 2010; Otoo, 2014; Snee, 2013). My field notes reflect the tension I felt during these conversations. At the time, I was unemployed and using my scholarship payments to support myself through fieldwork. There were bills I had to help my husband pay in Australia. By Australian standards, I was living below the poverty line. In Zimbabwe, I was relatively wealthy. In these conversations I tried to emphasise to my research participants that the cost of living was much higher in Australia, and that I could not afford to sponsor their children. This complexity is something I continue to navigate, as I reflect on my experiences during my fieldwork. In moments of reflection, I ask myself whether the response I gave was correct, whether I should have tried harder to find someone to fund the projects the participants mentioned, or whether I should have given them some of my own money. There are no easy answers in the messiness of navigating issues of privilege and race.

## Summary

Matopo School is one of many schools in the Matabeleland North region that receives tourist visits. The visits are frequent and there are several companies that visit the school. The tour has multiple configurations but they all have similar components: greeting, dancing/singing, school history, and school tour. Matopo School has some flexibility in its funding arrangements and in the way it runs as it is a satellite school, not a government school.

## Notes

1 This percentage does not include the $200/US cash donation, as this is provided to the school in cash on each visit and used for small purchases or emergencies.
2 These percentages do not total 100%, as the rest is from personal/other funding not related to any of the guests associated with the companies listed.

## References

Abebe, T. (2009). Multiple methods, complex dilemmas: Negotiating socio-ethical spaces in participatory research with disadvantaged children. *Children's Geographies*, *7*(4), 451–465. https://doi.org/10.1080/14733280903234519

Ahmed, S. (2007). A phenomenology of whiteness. *Feminist Theory*, *8*(2), 149–168. https://doi.org/10.1177/1464700107078139

Bonnett, A. (1997). Geography, 'race' and whiteness: Invisible traditions and current challenges. *Area*, *29*(3), 193–199.

Brandt, F., & Josefsson, J. (2017). Sexuality and power on South African game farms; reflections on positionality and emotions in ethnographic research. *Emotion, Space and Society*, *23*, 26–32. https://doi.org/10.1016/j.emospa.2017.02.004

Cabrera, N. L. (2016). When racism and masculinity collide: Some methodological considerations from a Man of Colour studying Whiteness. *Whiteness and Education*, *1*(1), 15–25. https://doi.org/10.1080/13613324.2015.1122662

Chilisa, B. (2012). *Indigenous research methodologies*. SAGE Publications Inc.

Chilufya, A., Hughes, E., & Scheyvens, R. (2019). Tourists and community development: Corporate social responsibility or tourist social responsibility? *Journal of Sustainable Tourism*, *27*(10), 1513–1529.

Chitiyo, G., & Chitiyo, M. (2009). The impact of the HIV/AIDS and economic crises on orphans and other vulnerable children in Zimbabwe. *Childhood Education*, *85*(6), 347–351.

Chitiyo, G., Chitiyo, M., Rumano, M., Ametepee, L. K., & Chitiyo, J. (2010). Zimbabwe education system: Emerging challenges and the implications for policy and research. *Journal of Global Intelligence and Policy*, *3*(3), 35–42.

Clausen, H. B. (2019). NGOs, tourism and development. In R. Sharpley & D. Harrison (Eds.), *A research agenda for tourism and development* (pp. 71–87). Edward Elgar Publishing Limited.

Dube, B., Maphosa, V., & Jita, T. (2021). Mobile learning app in the COVID-19 crisis among high school students in Zimbabwe: Construction of new learning. *Multicultural Education*, *7*(3), 251–258.

Ellingson, L. L. (2017). *Embodiment in qualitative research*. Routledge.

Ellsworth, E. (1989). Why doesn't this feel empowering? Working through the repressive myths of critical pedagogy. *Harvard Educational Review*, *59*(3), 297–324.

Faria, C., & Mollett, S. (2014). Critical feminist reflexivity and the politics of whiteness in the 'field'. *Gender, Place & Culture*, *23*(1), 79–93. https://doi.org/10.1080/0966369x.2014.958065

Fine, G. A. (1993). Ten lies of ethnography: Moral dilemmas of field research. *Journal of Contemporary Ethnography*, *22*(3), 267–294.

Fine, M. (1994). Working the hyphens: Reinventing Self and Other in qualitative research. In N. K. Denzin & Y. S. Lincoln (Eds.), *Handbook of Qualitative research* (pp. 70–82). SAGE Publications Inc.

Hammersley, M. (2017). What is ethnography? Can it survive? Should it? *Ethnography and Education*, *13*(1), 1–17. https://doi.org/10.1080/17457823.2017.1298458

Hanson, R., & Richards, P. (2017). Sexual Harassment and the Construction of Ethnographic Knowledge. *Sociological Forum*, *32*(3), 587–609. https://doi.org/10.1111/socf.12350

Harper, M., & Smithers, K. (2023). Fieldwork from A–Z? Exploring shifting identities in doctoral research in Australia and Zimbabwe. *The Australian Educational Researcher*. https://doi.org/10.1007/s13384-023-00649-1

Hove, M., & Ndawana, E. (2019). Education provision in the midst of a crisis: The Zimbabwean experience after 1999. *Journal of Peace Education*, *16*(2), 215–216. https://doi.org/10.1080/17400201.2019.1627663

Lozanski, K. (2010). Defining 'real India': Representations of authenticity in independent travel. *Social Identities*, *16*(6), 741–762. https://doi.org/10.1080/13504630.2010.524781

Madziyire, G. T. (2015). *Evaluating the impact of philanthropic activities in public high schools in Mutasa district, Zimbabwe: An educational management perspective* [University of South Africa]. https://core.ac.uk/download/pdf/43177311.pdf

Ministry of Primary and Secondary Education. (2017). *Annual education statistics profile Matabeleland North Province 2017*. http://mopse.co.zw/sites/default/files/public/downloads/2017_Matabeleland_North_Province_Statistics_Profile.pdf

Ministry of Primary and Secondary Education. (2019). *2018 Primary and secondary education statistics report*. Ministry of Primary and Secondary Education. Retrieved November 10 from http://mopse.co.zw/sites/default/files/public/downloads/2018%20Annual%20Statistics%20Report.pdf

Moreton-Robinson, A. (2002). *Talkin' up to the white woman: Aboriginal women and feminism*. University of Queensland Press.

Mostafanezhad, M. (2014). *Volunteer tourism: Popular humanitarianism in neoliberal times*. Ashgate Publishing Limited.

Mutana, S., Chipfuva, T., & Muchenje, B. (2013). Is tourism in Zimbabwe developing with the poor in mind? Assessing the pro-poor involvement of tourism operators located near rural areas in Zimbabwe. *Asian Social Science*, *9*(5), 154–161.

Mutana, S., & Zinyemba, A. Z. (2013). Rebranding the Zimbabwe tourism product: A case for innovate packaging. *International Journal of Advanced Research in Management and Social Sciences*, *2*(4), 95–105.

Mutanga, C. N., Vengesayi, S., Gandiwa, E., & Muboko, N. (2015). Community perceptions of wildlife conservation and tourism: A case study of communities adjacent to four protected areas in Zimbabwe. *Tropical Conservation Science*, *8*(2), 564–582.

Otoo, F. E. (2014). Constraints of international volunteering: A study of volunteer tourists to Ghana. *Tourism Management Perspectives*, *12*, 15–22. https://doi.org/10.1016/j.tmp.2014.07.002

Pollard, A. (2009). Field of screams: Difficulty and ethnographic fieldwork. *Anthropology Matters Journal*, *11*(2).

Scarth, A., & Novelli, M. (2019). Travel philanthropy and development. In R. Sharpley & D. Harrison (Eds.), *A research agenda for tourism and development* (pp. 88–109). Edward Elgar Publishing Limited.

Snee, H. (2013). Framing the Other: Cosmopolitanism and the representation of difference in gap year narratives. *The British Journal of Sociology*, *64*(1), 142–162. https://doi.org/10.1111/1468-4446.12010

Spivak, G. C. (1998). Can the subaltern speak? In C. Nelson & L. Grossberg (Eds.), *Marxism and the Interpretation of Culture* (pp. 271–313). University of Illinois Press.

Tawodzera, M. C., & Themane, M. (2019). Schooling experiences of children left behind in Zimbabwe by emigrating parents: Implications for inclusive education. *South African Journal of Education*, *39*(Supplement 1), S1–S13. https://doi.org/10.15700/saje.v39ns1a1797

Walford, G. (2009). The practice of writing ethnographic fieldnotes. *Ethnography and Education, 4*(2), 117–130.

Wondirad, A., Tolkach, D., & King, B. (2020). NGOs in ecotourism: Patrons of sustainability or neo-colonial agents? Evidence from Africa. *Tourism Recreation Research, 45*(2), 144–160. https://doi.org/10.1080/02508281.2019.1675269

ZIMSTAT. (2020). *Poverty and social impacts of COVID-19: Results from the rapid PICES phone survey data.* http://www.zimstat.co.zw/wp-content/uploads/publications/Income/Finance/RAPID_PICES_Wave1.pdf

# 4    The white saviour complex

The image of the 'African' child is somewhat inescapable for those in the Minority World. As a child raised in the 1990s and 2000s, I participated in an event called 'the 40-hour famine' – a global event that continues to this day. When I was a teenager, the purpose of the 40-hour famine was to exist solely on water and barley sugar lollies for forty hours, whilst raising money for the charity 'World Vision'. Within this challenge, practices of starvation were supposed to enable the individual to appease some guilt surrounding structural inequality and to practice empathy towards those in other countries. The year I participated, my friends and I organised a sleepover to support each other as we undertook this task. It was a weekend of relaxing, gossiping, and fun – all within the confines of a house filled with food and other luxuries. Although we knew the purpose of the challenge was to raise awareness and funds, most of the weekend was subsumed with fun rather than any sort of empathy building or reflection on our own privilege. By all accounts, the main purpose of the challenge was somewhat obscured by the activities we participated in as a group of teenagers having fun on a weekend. Arguably, as teenagers growing up in Australia, we were not well equipped to reflect in this way, and we were not well educated in the causes of global inequality. This also prevented us from having any deeper learning as a result of the challenge.

In the intervening years, there has been growing critique of an event that encourages voluntary starvation. As more awareness is brought to activities such as the 40-hour famine, the organisers have shifted the focus slightly. For example, now participants are encouraged to give up caffeine, take away food, or sugar for 40 hours or 40 days instead of forgoing food entirely. There has also been an increasing number of other charitable challenges aimed at fundraising and awareness raising. These charities seem to be ubiquitous in the Minority World. In Australia, there are 'do it in a dress' and 'ration' challenges that encourage people to donate money and to engage in acts that promote understanding and empathy.

DOI: 10.4324/9781003407980-4

The 40-hour challenge is not World Vision's only endeavour. They also facilitate sponsorship of children. Worldwide, an estimated 9 million children have been supported through various international sponsorship programmes (Wydick et al., 2013). Child sponsorship programmes usually require a monthly commitment by the sponsor and provide income for school fees or other community programmes. The sponsor is provided with a picture of the child and encouraged to write them letters. The flow of these programmes is usually Minority World to Majority World. These programmes are not well documented and there is limited evaluation of their efficacy and impact on the communities involved. At their heart, these programmes rely on communication between the sponsor and child, enabling the sponsor to learn more about the sponsored. Sometimes the sponsor will visit the child in their home country and may send extra financial support or gifts. Some sponsors display pictures of the child in their home.

The advertising of sponsorship programmes relies on the image of the African child – one of poverty. The images invoked are described by Sheridan et al. (2019) as the 'Black Baby phenomenon' where at the crux of the advertising is 'seeing Africans as black babies to be saved by donating pennies to religious charities' (p. 859). In this book I use the idea of the 'poor-but-happy' child to frame discussions of the image of African children. The sponsorship programmes develop a donor-recipient relation that mirrors broader geographies of care and compassion (Rabbitts, 2012). Bornstein (2001) discusses World Vision's interventions in Zimbabwe and identifies the tension inherent in sponsorship programmes, 'the transcendent aspirations of philanthropic practice not only fail to transcend difference – they may magnify and reconstitute economic disparity' (p. 615). Child sponsorship schemes, and other philanthropic endeavours, enable those in the Minority World to feel as if they are contributing to the alleviation of structural inequalities by providing education or economic support to an individual child. These philanthropic activities of the individual cannot be unravelled from the bigger concept of 'development'.

Development is a Minority World construct, which posits that all countries can be measured on a scale of progression, from developing to developed, with Minority World nations being the most 'developed'. This idea privileges Minority World ways of knowing and doing and is a problematic construct that fails to mask older constructs which sought to divide such as First/Third World. The scale of progression, which characterises Development, allows for the introduction of an insidious regime of truth: that of the white saviour. The 'white saviour complex' is an insidious discourse which has colonial origins and currently presents itself in development and volunteer tourism. The white saviour positions themselves as the saviour of the 'other'.

The history of Development, the concept of the 'poor-but-happy' child, and the white saviour complex permeate interactions in tourism. These three concepts, although deeply problematic, are integral to school tourism as they are a conceptual basis for the assumptions made by tourists when they arrive at the school. Further, in both tourism and volunteer tourism, the 'poor-but-happy' child is an image that romanticises poverty by justifying a lack of material assets. That is, poverty can be excused as the people are 'happy'. The image of the poor-but-happy child constitutes children in the Majority World in a manner which allows tourists to excuse the poverty they witness.

## Development

Largely, the idea of becoming 'Developed' is a Minority World invention, with the first widely accepted understandings of development originating from modernisation theory, in the wake of the Second World War (Mowforth & Munt, 2009). These ideas relied on the proliferation of the theory in which there is a dichotomy between 'traditional' under-developed societies and the 'modern' developed society. This is part of the deficit model of development, supported by the use of economic statistics, which posits that some societies have developed while others are less advanced (Bandyopadhyay & Patil, 2017; Jackson, 2013). In practice the dichotomy was conceptualised as a scale through which countries could move towards the category of 'developed' (Mowforth & Munt, 2009).

Development is many things at once: it is a discourse, a marker on which success is measured, a field of studies, and a field of employment (Mostafanezhad, 2014b; Mowforth & Munt, 2009). The multiplicity makes a discussion of 'development' tricky. The notion of Development is a hark back to colonial ideas of progress and progression towards a Minority World ideal, and as such it is a loaded and problematic term. Development has moved through different phases, largely driven by economic and political contexts (for a concise examination see Mowforth & Munt, 2009, p. 33). Grosfoguel (2007) captures the continuous episteme of marking the Other as inferior across time,

> We went from the sixteenth-century characterization of 'people without writing' to the eighteenth century and nineteenth century of 'people without history', to the twentieth-century characterization of 'people without development' and more recently, to the early twenty-first century of 'people without democracy'.
>
> (p. 214)

An in-depth tracing of Development, including its knowledge formations and the technical and administrative apparatus through which power relations govern conditions of possibility, has been explored elsewhere (see Escobar, 1984, 1995/2012; Ndlovu-Gatsheni, 2013).

Discourses of development are rooted in Eurocentric models of thinking but began in earnest as countries commenced the process of decolonisation (Mowforth & Munt, 2009). The discourse of Development emerged in the late 1940s following American President Truman's inaugural address, in which he introduced 'point 4': advancement of underdeveloped areas of the world (Hickel, 2017; Mowforth & Munt, 2009; Moyo, 2009). The concept of 'developing' those who were underprivileged spread into European imaginaries and became embedded in global discourses.

Mowforth and Munt (2009) identify development as having grown since Truman's presidential address and,

> Since that time, development, above all else, has signalled the need to Escape the undignified confines of underdevelopment... Development may be best understood foremost as a hegemonic discourse that originates from, and is largely fashioned by, First World dominated global institutions, governments, agencies and academe.
>
> (pp. 31–32)

In the discourses of Development all countries are, usually without problematisation, considered on their progress towards becoming 'developed'. There is little recognition by Minority World countries of the interlinked histories which influence present inequalities in Majority World countries (Kothari, 2006b; Spivak, 1998). Everyday language used in the Minority World is wrapped in these terms, for example, the term 'Third World', coined in 1952, was initially used by Alfred Suavy to distinguish political systems that were different to free-market capitalism and socialism (Baptista, 2011; Tomlinson, 2003). The term 'third world' became, however, a euphemism for countries that were classified as 'underdeveloped' and not sufficiently progressed towards a Minority World ideal.

At a similar time to the introduction of the term 'Third World', leaders of 44 nations met at Bretton-Woods, United States of America (Moyo, 2009; Peet, 2009). It was at this meeting that three economic institutions responsible for governing the international economy were developed: the International Monetary Fund (IMF); the International Bank for Reconstruction and Development, which later became known as the World Bank; and, the International Trade Organisation, which later became the World Trade Organisation (Peet, 2009). These policies were put in place as a reactionary measure to the inter-war period, with a function of the economic institutions being to maintain a collective responsibility for international monetary order (Peet, 2009).

On a world-economy scale, conceptualisations of countries on some sort of progression towards 'development' ignores the long history of 'underdeveloped' countries being used by Minority World countries as sources of cheap resources and labour (Spivak, 1998; Wallerstein, 2000). Further, it ignores the continuing economic, social, and cultural legacies of colonisation, deemed the 'colonial power matrix' by decolonial scholars (Grosfoguel, 2007; Quijano, 2010). As countries were decolonising and becoming independent, the former colonisers did not necessarily lose economic and political powers. There were deliberate and active strategies put in place by the colonial powers to continue to rob post-colonial countries of their resources, such as precious metals. For example, as detailed in Chapter 2, the World Bank and the IMF encouraged open-market economic principles whilst discouraging socialist economic practices; they were not necessarily value-neutral institutions.

Some argue that concerning economic global structures, we should be discussing the neo-colonial present, rather than the post-colonial present (Wijesinghe et al., 2019). Moyo (2009) argues that development as a tool for neocolonisation began in tandem with decolonisation efforts,

> For the West, aid became a means by which Britain and France combined their new-found altruism with a hefty dollop of self-interest – maintaining strategic geopolitical holds. For the US, aid became the tool of another political contest – the Cold War.
>
> (p. 27)

Neocolonisation, through development, is evident in the measures put in place by Minority World economic organisations such as the World Bank, UNESCO, and so on, in which countries are compared based on statistics and averages (Escobar, 1984, 1995/2012; Grosfoguel, 2007). These statistics give rise to a whole range of discourses in which countries are measured and declared as having met or not met the Minority World ideals of 'developed'. Paradoxically, conforming to Minority World requests and interventions is almost futile, as Majority World countries will be always marked as Other by Minority World knowledge formations in the current colonial power matrix that privileges Minority World epistemologies (Ndlovu-Gatsheni, 2013). By providing wealth and knowledge to those less privileged, it allows former colonial powers to develop a narrative of becoming 'benevolent caregivers' who will 'fix' the problems of those who are less developed (Durrheim et al., 2011; Wale & Foster, 2007).

As per the idea that a country can become sufficiently 'developed', global aid is justified as further developing the 'underdeveloped' countries by providing the things they are 'missing' (Jackson, 2013). The colonial powers created a network of global 'haves' and global 'have nots' through the deficiency model, which first distinguished the 'third world' and 'first world' (Jackson, 2013; Ndlovu-Gatsheni, 2013). This binary perpetuates unequal power relations as people in the Minority World feel they can intervene in the Majority World in the name of development (Escobar, 1995/2012; Ndlovu-Gatsheni, 2013). Further, the notion of all countries being considered on a scale of developed or undeveloped obscures the histories of colonisation and the inequalities which result from years of oppression due to colonisation (Matthews, 2017; Spivak, 1998). As McEwan (2001) identifies, 'development is about power – its operations, its geographies, its highly uneven distribution and strategies for achieving it' (p. 93). The system of 'under-development' has a relatively short history but is an example of the continuing colonial discourses of a marked 'other'; with negative representations of the African continent once propagated by travellers, missionaries, and anthropologists, now being furthered by NGOs and the media (Ndlovu-Gatsheni, 2013).

Development, white supremacy, and colonial legacies have interlinked histories (Dirlik, 1994; Kobayashi, 2003; Kobayashi & Peake, 2000, 2007). Through an examination of economic and statistical markers, the Majority World is marked by the Minority World as unsuitable for controlling their own policies and economic futures. The discourses of Development are enabled by the use of statistics and economic measures which present as self-evident statistical 'truths' used to measure progression towards development (Escobar, 1995/2012; Kothari, 2006b). The self-evident 'truths' about modernity mask the racial history of development which has shaped the unequal distribution of wealth through economic interventions from the IMF and other economic institutions (Kothari, 2006a). There have been significant interventions in Zimbabwe by the World Bank and the IMF in the name of economic development, such as the move from socialist policies to a free market economy as discussed in Chapter 2. These interventions forced structural adjustments to economies and encouraged privatisation in the name of development but were widely unsuccessful – not just in Zimbabwe (Gardner, 2016; Mowforth & Munt, 2009).

Mostafanezhad (2014a) argues that in recent times there has been a move from Development as a practice, to Development as a discourse. I follow the argument of Ndlovu-Gatsheni (2013), which is built from Escobar (1995/2012), in understanding Development as a discourse. By understanding Development as a discourse, we can understand it as part of the making of social reality. Further, Ndlovu-Gatsheni (2013) argues that 'it is a form of articulation of knowledge and power that cannot simply be depoliticised into mere life issues of poverty and depravity' (p. 106).

With the rise of statistical measure of success towards 'development' there has been a rise of NGOs and organisations who work in the name of development. White supremacy is embedded in policy decisions and the whiteness of the workers is one of the most 'visible' signs of difference within the work of development (Conran, 2011). In fact, Mason (2016) identifies that 'development is always already set within a frame of racism (where the project of development is historically and contemporarily a racial project)' (p. 830).

Whilst whiteness is a very 'visible' issue within development, there is very little attention paid to the effects of race in the work of development (Pailey, 2019). Faria and Mollett (2014) in particular, argue for a 'whitening' of the field of development that raises awareness of the whiteness of development. They argue this can be achieved through the use of postcolonial feminism, with a 'deliberate tracing of how coloniality and whiteness continue to shape knowledge production' (p. 81). Kapoor (2004) also takes development to task, suggesting that 'encounters with, and representations of, our "subjects" are therefore coded or framed in terms of an us/them dichotomy in which "we" aid/develop/civilise/empower "them"' (p. 629). These conceptualisations of both whiteness and Development are important in the context of school tours as they inevitability frame the way in which development through tourism is always-already defined and constructed.

## Developmentourism and education: problems of naming

Development *through* tourism has a relatively short history, with the advent of the so-called 'moral' tourisms, such as volunteer tourism and philanthropic tourism, occurring more recently as a response to criticisms of mass tourism.

A term coined by Baptista (2011) is developmentourism, which acknowledges the ways in which development and tourism are combined in attempts to create more 'moral', community-based, alternative tourisms. By blending the two terms of development and tourism Baptista (2011) argues for an envisaging of this type of tourism as both encapsulating the discourses associated with development, and the neoliberal properties of commodification:

> More precisely, the representations of 'donor' and 'tourist' in Mbueca and Canhane are a reflex of the market, ruled by tourism and 'development' principles, where the members of the 'communities' act, as product, producers and sellers. The cases of Mbueca and Canhane, and the dedifferentiated 'tourist', thus inform the wider context in which they operate, and demonstrate how 'development' in the 'Third World' has become part of the entertainment industry, which in turn reflects the incessant creation of new consumer markets and products induced by the expansion of neo-liberalism.
>
> (p. 664)

Developmentourism, then, is reflective of the marketisation and commodification of 'development'. There are a range of different tourism(s) involved with schools, in a range of capacities, at the heart of the engagements are donations and philanthropy to schools based as they are in discourses of development and whiteness.

As part of these new tourisms, schools are becoming tourism destinations and becoming part of the tourism product (I have explored this in depth elsewhere, see Smithers & Ailwood, 2022). Table 4.1 details the variety of tourism niches that include visits or relationships to schools. Some tourisms include different types of engagement, making the definitions blurry.

The demarcation of these tourism into increasingly smaller niches is somewhat problematic, as it makes a scoping of the extent of school involvement in tourism difficult. There is also the possibility that by demarcating into infinite niches, the overarching development discourses are obscured – ensuring the ethical dimensions and deeply embedded neo-colonial narratives of these tourisms remain uninterrogated. I now discuss some of these tourisms, highlighting the interplay of terms and showing how many of these labels are types of tourism that could be considered within Baptista's (2011) definition of developmentourism.

Philanthropic tourism involves providing gifts or donations at some point in the travel journey or travelling with the purpose of participating in a philanthropic activity (Scarth & Novelli, 2019). The term encompasses a range of countries and tourism ventures, with activities ranging from charity car trips to visits to view the results of philanthropic donations (Lacey et al., 2016; Scarth & Novelli, 2019). Philanthropic tourism can be distinguished from other forms of tourism through the act of visiting a charitable project, or specifically travelling for the purpose of philanthropy. Scarth and Novelli (2019) identify that

*Table 4.1* Types of tourism with relationships to schools

| Type of tourism | Involvement with schools | Name of tourism |
| --- | --- | --- |
| Cultural tours | School visit as part of larger cultural tour | • Community-based tourism<br>• Cultural tourism<br>• Developmentourism<br>• Orphanage tourism<br>• Pro-poor tourism<br>• Rural tourism<br>• Slum tourism<br>• Sustainable tourism<br>• Village tourism<br>• Byproduct of wildlife tourism<br>• Mass tourism |
| Tourism involving teaching | Volunteer/paid teacher immersed in school as part of experience (2 weeks – up to a year) with additional tourism activities included | • Gap year tourism<br>• Philanthropic tourism<br>• Teaching English as a Foreign Language (TEFOL) tourism<br>• Volunteer tourism |
| Visit to a school as part of philanthropic venture | School visit is the main purpose of the tourism (a few hours up to a year) with additional tourism activities sometimes included | • Mission tourism<br>• Philanthropic tourism<br>• Pro-poor tourism<br>• Volunteer tourism<br>• Corporate social responsibility byproduct |
| Varied – usually mass tourism ventures | Donation to school as byproduct of tourism – no visit | • Community-based tourism<br>• Ecotourism<br>• Mass tourism<br>• Pro-poor tourism<br>• Sustainable tourism<br>• Village tourism<br>• Wildlife tourism |

there is no consistent definition for philanthropic tourism, and this seems to play out in the variety of studies that refer to what could be considered philanthropic tourism, but using a different term. Lacey et al. (2016) examined philanthropic tourism in a school in Kenya which was mostly funded through a private trust and financed by donors. The school closed due to mismanagement and allegations made by donors against the board of trustees. In another study, the term 'philanthrotourism' is used to describe tourism with the purpose of visiting a recipient school (Koot & Fletcher, 2021). School tours which are conducted for the purpose of philanthropy are also included under the broad umbrella of 'corporate social responsibility' in Zambia (Chilufya et al., 2019).

Burns and Barrie (2005) describe a school funded through sustainable ecotourism, in which donors visited a lodge nearby and subsequently funded the development of a school. One of the participants said the following:

He [Christine's husband, Ray] was speaking to one of the local guys, who actually happens to work for the lodge, who said 'you know, why don't you..' like as a joke, 'why don't you build her a classroom, she can come and teach here for us, if she really misses teaching so much', and Ray thought this was a great idea. So he had the classroom built and brought her back the next year for her birthday, she thought the trip was her birthday present which it wasn't, it was her classroom, she loved the idea ... [Christine] brings them [her Canadian business staff] to actually come and do, to come and work and they have adopted the project, so they feel like they own it, their own little world, that they own a piece of Africa, it's a great holiday experience for them.

(Burns & Barrie, 2005, p. 469)

The extract highlights the complexities of naming these types of tourism. What began as a visit to a sustainable lodge (ecotourism), becomes a donation (developmenttourism), which then becomes more regular tourist trips with other potential donors (philanthropic tourism). The extract also makes clear how deeply embedded colonial discourses of paternalism are, in which the donors feel they 'own a piece of Africa'.

Using the example of two communities in Mozambique, Baptista identified a school tour that was a larger component of a village tour. The village tour, and subsequent school tours, was used by residents to demonstrate the benefits of local hotels funding development projects. This type of tourism is often also called community-based tourism; when a hotel/lodge/business provide both jobs and infrastructure to the community in which it is based (Baptista, 2011). However, the act of visiting a village nearby a hotel/lodge is also described by Chilufya et al. (2019) as 'corporate social responsibility'. To complicate matters, Scarth and Novelli (2019) identify visiting a school as part of a mass tourism package as a type of travel philanthropy that results from the 'compassionate by-products of a wildlife safari' (p. 90).

It seems that a school tour can sit under many niche tourism definitions, but very few of these studies identify the school tour as a specific type of tourism activity that has its own set of problems and ethical challenges. As I will discuss in this chapter and in later chapters, this is problematic, as tourism products like school tours rely on two images to sustain public interest: the 'poor-but-happy' image and the white saviour complex. These images, when they remain unquestioned and unchallenged can have damaging implications for host communities.

## The white saviour complex

The 'white saviour complex' is a term which recognises the discursive framing of white people 'saving' non-white people, usually in countries other than their own (Bandyopadhyay, 2019; Bandyopadhyay & Patil, 2017; Bex & Craps, 2016; Cammarota, 2011; Hughey, 2010). The white saviour discourse is infused with remnants of colonial paternalism (Bell, 2013; Everingham & Motta, 2020). The white saviour image is almost inescapable; it is embedded in media, literature, tourism advertising, and in the resulting holiday photographs from such holidays.

Celebrities and other public figures engage in humanitarian and development projects as tools of visible philanthropy. English royalty, including Prince Harry and Prince William, have conducted volunteer work in developing countries (Callanan & Thomas, 2005). More broadly, celebrities such as Madonna, Angelina Jolie, and Bono (as examined by Bell, 2013) all appropriate the image of 'doing' development into something to be admired (Hickel, 2017). The increase of celebrity efforts in development and humanitarian projects is an indication of the ways in which development and commodification are increasingly combined to create the image of a virtuous saviour of those marked as 'other' (Baptista, 2012). It could be argued that Bob Geldof, 'pushed this model for celebrity-led reaction to such an extent that celebrities are now a vital part of the "aid" industry' (Barker, 2013, p. 97).

The circulation of images of celebrities undertaking 'virtuous' acts perpetuate the white saviour discourse (Baptista, 2011, 2012). Coupled with the rise in alternative tourisms which encourage the image of a 'virtuous tourist' to exist, Cameron and Haanstra (2008) identify that development has had an image renovation,

> Past associations of the supporters of development causes in the North with socks, sandals and granola are replaced with a new imaginary of high heels, tight dresses and martinis, that is, a sophisticated, affluent, cosmopolitan and sexy Northern donor.
>
> (p. 1476)

The image renovation is beyond just celebrities. In a study by Burns and Barrie (2005) a participant labelled their own development work as 'sexy' and then linked it to the affective feel created by the control of nature, which is inherent in safari tourism,

> It is quite a sexy sort of area to actually be involved with, it's nice that you are feeding starving people ... Africa's like the last big frontier, Africa is very exciting and you take people on safari and I promise you there is no ways they cannot actually get excited about it.
>
> (p. 476)

Not only has development become 'sexy', but it has also increasing celebrity influence and support. I will later argue that the school teachers and the tourists who visited the school in this study were both agents of, and subjected to, a governing order about what both development and the virtuous tourist should look like. The white saviour discourse is produced through discourses of whiteness, colonialism, and development, in which the noble white saviour works to 'save' those on the 'dark continent'.

Baptista (2012, 2014) critiqued the commodification of development and the increasing influence of entertainment and media on the image of those who 'help'. As part of the commodification of development, a number of tourists seek experiences that will minimise their perceived impact on the local population (Nyahunzvi, 2013). One example is volunteer tourism in which tourists pay to volunteer in another country, with Bandyopadhyay and Patil (2017) arguing it is an 'exemplar' of the white saviour discourse (p. 646). Volunteer tourism is often dubbed an 'ethical' form of tourism – if only by its supporters (Butcher, 2017). It is a form of tourism which aims to satisfy the need for participants to 'give back' while simultaneously appearing to provide an alternate outlet to traditional mass tourism.

Despite recent research popularity, there is no universal definition of volunteer tourism or volunteer tourists (Callanan & Thomas, 2005; Stainton, 2016; Wearing et al., 2017), with Taplin et al. (2014) identifying twenty-one different types of volunteer tourism on the market. Predominantly volunteer tourists are 18–25, reside in the Minority World, and want to volunteer in the Majority World with the most popular destination being the continent of Africa (Butcher & Smith, 2010; Kontogeorgopoulos, 2017; Proyrungroj, 2017). The flow of Minority World tourists to Majority World destinations in search of 'helping' suggests that volunteer tourism is enmeshed deeply with discourses of the white saviour and a particular image of the needy 'Other'.

There is no doubt that volunteer tourism is embedded with discourses of the white saviour. In the Minority World, historically, an image of Africa as primitive and backwards helped to justify the 'civilising' mission and the exploitation of both African people and the African landscape (Mudimbe, 1988). These discourses have very real consequences for the way in which the Other is constructed in Minority World imaginaries in the present. For example, Nyahunzvi (2013) analysed advertising materials of 25 organisations that offer volunteer experiences in Zimbabwe, and most used the rhetoric of making a 'real difference' to the lives of 'poverty-stricken others'. Henry (2020) identifies that tourism scholars need to consider volunteer tourism as a racialised project, 'to theorize voluntourism as a racialised encounter, scholars need to consider its basic premise, that a world out there exists and is available for certain people to enter and "improve"' (p. 327). Volunteer tourism is a racialised project, however scant research attention has been paid to the role that whiteness and the white saviour complex play in volunteer tourism (Bandyopadhyay, 2019; Bandyopadhyay & Patil, 2017). Similar to school tours, volunteer tourism is framed through discourses of the white saviour and Development (Bandyopadhyay, 2019; Bandyopadhyay & Patil, 2017). The school tour is not volunteer tourism, but it has striking similarities. In particular, the mix of leisure activities, cultural activities, and philanthropic activities is reminiscent of volunteer tourism itineraries. I will argue in later chapters that the school tour is imbued with discourses of Development and the white saviour.

Volunteer tourism needs to be problematised not just in terms of the white saviour discourses but also in terms of child-safety and the impact on host communities. These serious issues are often not considered by volunteers or by tourists, as the discourses of care and paternalism constitute the Majority World as in need of saving. Wright (2014) identifies the ethical implications of volunteer tourism in schools, in which one research participant highlighted the negative effects of tourism in schools stating, 'they will just be taught by random people, probably the same thing over and over again' (p. 245). Short-term teachers were found by Bargeman et al. (2018) to have ignored the Ghanaian curriculum, instead choosing to teach content of their own interest, meaning the local teachers had to fill in the gaps in learning once the volunteers had left. It is concerning that children's education is compromised by young volunteers from the Minority World who believe they are helping.

Interestingly, little is known about the scope and impact of volunteer tourism activities in schools – given its relative popularity as a volunteer tourist activity. Jakubiak (2016), for example, examined 'English language voluntourism' by interviewing participants in a number of programmes in Costa Rica. Bargeman et al. (2018) examined the impact of volunteer tourism on a Ghanaian school. They found that often there was a mismatch between the expectations of volunteers and the expectations of host teachers. This resulted in confusion, volunteers teaching outside of the prescribed curriculum and a loss of key learning time for students. Thompson and Taheri (2020) argue that hosts of volunteer tourists are often constrained by economic conditions which render them unable to speak up regarding the disruptive influence of volunteer tourists on schools. It appears then that although little is known about volunteer tourists in schools – what is known is not positive. They interrupt the delivery of curriculum, often re-teach the same content, and the host communities feel they cannot speak back against the white saviour in action.

To add complexity to the issue, the white saviour discourse is not only permeating globally, but on a local level within the continent of Africa. There are significant connotations of this discourse, detailed by Durrheim et al. (2011):

> The discourses blame the victims of colonial and capitalist exploitation by identifying African people as the cause of their own problems. At the same time the discourse exonerates the white capitalists and administrators, identifying them as knowledgeable guardians and good, caring people.
>
> (p. 102)

They continue on to say,

> With the wealth they accumulated the colonialists could do 'good works' such as building hospitals, churches and schools. Their self-perceived superiority was thus double loaded: not only were they more civilised than Africans to start with, but they became even nobler by doing such good things to develop Africa.
>
> (Durrheim et al., 2011, p. 109)

The discourse, which portrays black Africans as needing guardianship and protection, allows white and middle-class South Africans to adopt a 'philanthropic self-identity' in which they can take care of the black Africans who do not know what is good for them. The white saviour complex ignores, in part, the structural barriers which cause poverty by placing responsibility on the white middle class to support those in poverty (Wale & Foster, 2007). White Zimbabweans also take up these white saviour discourses, particularly in terms of paternalism displayed for their workers (Alexander, 2004; Hartnack, 2015; Misi, 2016; Pilossof, 2009; Suzuki, 2017, 2018; Tagwirei & de Kock, 2015).

Identity through the Other, and the idea of white identity building through the Other is perpetuated by discourses that suggest that individuals can, or should, save the Other, both children and adult. This discourse extends beyond just local white identities and permeates into the discourses surrounding humanitarianism and development (Mostafanezhad, 2013a, 2013b; Sensoy & Marshall, 2010). The white saviour complex and images of children go hand-in-hand.

## Dominant images of Majority World children

Having outlined Development and the white saviour as dominant discourses which govern tourism formations, I now explore one of the dominant images of Majority World children. There are competing images and of children, as constructed by the Minority World (Lahman, 2008). These competing discursive constructions of Majority World children include the image of the child-soldier (Martins, 2011), the impoverished child (Sinervo & Hill, 2011), the scheming child (Murris, 2013; Nieuwenhuys, 2013) and the 'poor-but-happy' child (Bandyopadhyay & Patil, 2017; Crossley, 2012; Wearing et al., 2018). Although there are many competing images of children, the most prevalent to this book is the poor-but-happy image, as the school actively worked to construct an image of happy, smiling children. I now explore the poor-but-happy construction and use volunteer tourism to exemplify the relationship between the poor-but-happy child and the white saviour complex.

The image of a poor-but-happy child permeates discourses associated with foreign aid and advertising (Mostafanezhad, 2014a). In the poor-but-happy discourse, people – mostly children – are depicted as impoverished but happy. This dominant discourse can be seen in accounts of overseas holidays, for example, people might say 'they had so little, but they appreciated everything they have' or 'the children were so poor, but they still smiled and played'. Further, in advertising this image is displayed in the form of children smiling at the camera or in depictions of children in schools (Rideout, 2011). The poor-but-happy discourse flattens the image of the child, rather than being a multifaceted human being, it reduces them to a smiling, dancing figure. This has links to the exotification of cultures in tourism ventures. For example, there have been some examinations of this discourse in relation to postcards (Sinervo & Hill, 2011; Sun & Yanjun, 2020).

The poor-but-happy discourse is paternalistic, as it suggests that those who do not have the material affluence of the Minority World are naïve because they do not know what their lives are missing. Further, it ignores the reality in which there is an existence of Minority World influence in the Majority World, such as the permeation of American popular culture through television and media. The poor-but-happy child paints a picture of innocent children who are oblivious to the world, fulfilling the destination image of an African lifestyle which is simpler and an escape from a consumerist lifestyle. In addition, the poor-but-happy discourse place the child and the tourist in an unequal relation of power, whereby the Minority World traveller can impart worldly advice or knowledge onto the child, without acknowledging the culture or knowledge the child could provide to the tourist. Any interactions, then, are most likely governed by paternalistic discourses which constitute the child as needy.

The poor-but-happy discourse is a 'discursive romanticisation of poverty' (Frazer & Waitt, 2016, p. 184) through which poverty is romanticised. The image of happy children allows for a sanitised image of poverty to be communicated in the Minority World. Mostafanezhad (2013a) argues children's role in humanitarianism 'highlights the pervasive linkage between the colonial iconography of childhood and savagery and the more recent emergence of the child as a symbol for universal human rights' (p. 331). There are serious issues associated with the poor-but-happy discourse. For instance, it implies that despite living in poverty, the children are happy with their place in the world system. In part, this means those in the Minority World can ignore the structural inequalities which influence the continuation of poverty (Frazer & Waitt, 2016). The poor-but-happy discourse enables the tourist to ignore the uneven socioeconomic position which enables them to pay for a cross-continental flight and undertake travel in another country, whilst gazing upon a child who is much poorer than themselves (Frazer & Waitt, 2016). Further, the poor-but-happy discourse enables the tourist to excuse the poverty they witness, as they see children who are 'happy'. In essence, it means that the tourist does not have to take action about the poverty they are seeing as they can justify that the children are happy with their place in life (Santoro, 2014).

Volunteer tourism perpetuates the poor-but-happy discourses as it enables young tourists from the Minority World to visit those in the Majority World and excuse the poverty they witness. There have been examinations of volunteer tourism and service-learning trips, highlighting the ways in which they perpetuate and encourage the discourse of the poor-but-happy child and the noble white saviour (Crossley, 2012; Frazer & Waitt, 2016; Judge, 2016; Santoro, 2014). Mostafanezhad (2013a) argues that volunteer tourism is imbued with parent-child imagery that was characteristic of colonisation; volunteer tourism usually relies on children in which they are constructed as dependent on those around them for protection. Other studies have found that volunteer tourism promotes feelings of 'love' and care for children in tourism destinations and that these feelings are representative of colonial discourses which continue in the present (Bandyopadhyay & Patil, 2017; Judge, 2015). I argue later in the book that these images are also embedded in school tours and imagery surrounding school tours.

**Summary**

Development and discourses of development are far reaching in the Minority World. They exist in images of Majority World children, but also in the prevalence of child sponsorship schemes, charitable events, and in images of celebrity humanitarianism. Development discourses privilege Minority World ways of knowing and doing, whilst ignoring the damaging colonial histories for most of the Majority World. Wrapped in these ideas of development is the idea of global aid, which is suggested as a potential solution to 'under-development'. Geographies of care and compassion are mirrored in the relations between donor and recipient. These relations shape the conditions of possibility for the shaping of school tours but also the potential reactions by participants.

There exist several iterations of tourism in schools, which rely on development discourses as an underpinning ideal. These include school tours, volunteer tourism, slum tourism, and more. Although the tourism industry has multiple labels for these types of tourism and appears to splinter them into multiple categories, they can all be considered as variants of the commercialisation of development, termed by Baptista as 'developmentourism'. These tourisms seem to be largely taken for granted by the tourism industry as a sidenote to other bigger industries. For example, school tours have been described as compassionate byproduct of wildlife tourism, and corporate social responsibility is most often considered in tandem with the profits or public perceptions of businesses.

The white saviour complex is a damaging and problematic trope which is furthered through celebrity humanitarianism and various tourism ventures. In the advertising and continuation of these types of activities, the idea of someone from the Minority World 'saving' the marked Other from the Majority World is normalised and remains relatively unquestioned.

I have shown there are multiple and competing images of Majority World children. The image I interrogate later is the poor-but-happy child, which is ubiquitous and problematic. It enables tourists to excuse extreme poverty and ignore the structural reasons for global inequalities. Further, in many examples of tourism the poor-but-happy child has become a spectacle to be viewed or 'saved' by the white saviour.

Combined, these two ideas of the white saviour and the poor-but-happy child – which are based in development discourses – work to enable conditions of possibility through which Minority World tourists feel they can seek out and witness other people's suffering. The images of the white saviour and the poor-but-happy child coalesce in volunteer tourism and development through tourism – particularly when these ventures include schools. I have mainly discussed volunteer tourism in this chapter, as it has recently experienced a hive of research activity surrounding it. Although there is very little research that discusses school tours, there is no doubt that school tours play into the same narratives of white saviour and poor-but-happy-child in their marketing and scope.

## References

Alexander, K. (2004). Orphans of the empire: An analysis of elements of white identity and ideology construction in Zimbabwe. In B. Raftopoulos & S. Tyrone (Eds.), *Zimbabwe: Injustice and Political Reconciliation* (pp. 193–212). Weaver Press.

Bandyopadhyay, R. (2019). Volunteer tourism and "The White Man's Burden": Globalization of suffering, white savior complex, religion and modernity. *Journal of Sustainable Tourism, 27*(3), 327–343. https://doi.org/10.1080/09669582.2019.1578361

Bandyopadhyay, R., & Patil, V. (2017). 'The white woman's burden' - the radicalized gendered politics of volunteer tourism. *Tourism Geographies, 19*(4), 644–657. https://doi.org/10.1080/14616688/2017.1298150

Baptista, J. A. (2011). The tourists of developmentourism– representations 'from below'. *Current Issues in Tourism, 14*(7), 651–667. https://doi.org/10.1080/13683500.2010.540314

Baptista, J. A. (2012). The virtuous tourist: Consumption, development, and nongovernmental governance in a Mozambican village. *American Anthropologist, 114*(4), 639–651. https://doi.org/10.1111/j.1458-1433.2012.01515.x

Baptista, J. A. (2014). Tourism moral imaginaries and the making of community. In N. B. Salazar & N. H. Graburn (Eds.), *Tourism Imaginaries: Anthropological approaches* (pp. 125–144). Berghahn Books.

Bargeman, B., Richards, G., & Govers, E. (2018). Volunteer tourism impacts in Ghana: A practice approach. *Current Issues in Tourism, 21*(13), 1486–1501. https://doi.org/10.1080/13683500.2015.1137277

Barker, M. J. (2013). Bob Geldof and the Aid Industry: "Do They Know it's Imperialism?". *Capitalism Nature Socialism, 25*(1), 96–110. https://doi.org/10.1080/10455752.2013.845586

Bell, K. M. (2013). Raising Africa? Celebrity and the rhetoric of the white saviour. *PORTAL Journal of Multidisciplinary International Studies, 10*(1). https://doi.org/10.5130/portal.v10i1.3185

Bex, S., & Craps, S. (2016). Humanitarianism, testimony, and the white savior industrial complex: What is what versus Kony 2012. *Cultural Critique, 92*, 32–56.

Bornstein, E. (2001). Child sponsorship, evangelism, and belonging in the work of World Vision Zimbabwe. *American Ethnologist, 28*(3), 595–622.

Burns, P. M., & Barrie, S. (2005). Race, space and 'Our own piece of Africa': Doing good in Luphisi village? *Journal of Sustainable Tourism, 13*(5), 478–485.

Butcher, J. (2017). The morality of mass tourism. In D. Harrison & R. Sharpley (Eds.), *Mass tourism in a small world* (pp. 16–40). CPI Group (UK) Ltd. https://doi.org/10.1079/978178068545.00000

Butcher, J., & Smith, P. (2010). 'Making a difference': Volunteer tourism and development. *Tourism Recreation Research, 35*(1), 27–36. https://doi.org/10/1080/02508281.2010.11081616

Callanan, M., & Thomas, S. (2005). Volunteer tourism: Deconstructing volunteer activities within a dynamic environment. In M. Novelli (Ed.), *Niche Tourism: Contemporary Issues, Trends and Cases* (pp. 192–209). Taylor & Francis.

Cameron, J., & Haanstra, A. (2008). Development made sexy: How it happened and what it means. *Third World Quarterly, 29*(8), 1475–1489.

Cammarota, J. (2011). Blindsided by the Avatar: White saviors and allies out of Hollywood and in education. *Pedagogy and Cultural Studies, 33*(3), 242–259. https://doi.org/10.1080/10714413.2011.585287

Chilufya, A., Hughes, E., & Scheyvens, R. (2019). Tourists and community development: Corporate social responsibility or tourist social responsibility? *Journal of Sustainable Tourism, 27*(10), 1513–1529.

Conran, M. (2011). They really love me! *Annals of Tourism Research, 38*(4), 1454–1473. https://doi.org/10.1016/j.annals.2011.03.014

Crossley, É. (2012). Poor but Happy: Volunteer Tourists' Encounters with Poverty. *Tourism Geographies, 14*(2), 235–253. https://doi.org/10.1080/14616688.2011.611165

Dirlik, A. (1994). The postcolonial aura: Third world criticism in the age of global capitalism. *Critical Inquiry, 20*(2), 328–356.

Durrheim, K., Mtose, X., & Brown, L. (2011). *Race trouble: Race, identity and inequality in post-apartheid South Africa.* Lexington Books.

Escobar, A. (1984). Discourse and power in development: Michel Foucault and the relevance of his work to the third world. *Alternatives, 10*(3), 377–400.

Escobar, A. (1995/2012). *Encountering development - The making and unmaking of the Third World.* Princeton University Press. (Paperback reissue 2012).

Everingham, P., & Motta, S. C. (2020). Decolonising the 'autonomy of affect' in volunteer tourism encounters. *Tourism Geographies.* https://doi.org/10.1080/14616688.2020.1713879

Faria, C., & Mollett, S. (2014). Critical feminist reflexivity and the politics of whiteness in the 'field'. *Gender, Place & Culture, 23*(1), 79–93. https://doi.org/10.1080/0966369x.2014.958065

Frazer, R., & Waitt, G. (2016). Pain, politics and volunteering in tourism studies. *Annals of Tourism Research, 57,* 176–189. https://doi.org/10.1016/j.annals.2016.01.001

Gardner, B. (2016). *Selling the Serengeti: The cultural politics of safari tourism.* The University of Georgia Press.

Grosfoguel, R. (2007). The epistemic decolonial turn: Beyond poltiical-economy paradigms. *Cultural Studies, 21*(2–3), 211–233. https://doi.org/10.1080/09502380601162514

Hartnack, A. (2015). Whiteness and shades of grey: Erasure, amnesia and the ethnography of Zimbabwe's whiteness. *Journal of Contemporary African Studies, 33*(2), 285–299. https://doi.org/10.1080/02589001.2013.873590

Henry, J. (2020). The cinematic pedagogies of underprepared teachers. *Teaching and Teacher Education, 89.* https://doi.org/10.1016/j.tate.2019.102990

Hickel, J. (2017). *The divide: A brief guide to global inequality and its solutions.* William Heinemann.

Hughey, M. W. (2010). The white savior film and reviewers' reception. *Symbolic Interaction, 33*(3), 475–496. https://doi.org/10.1525/si.2010.33.3.475

Jackson, L. (2013). They don't not want babies: Globalizing philosophy of education and the social imaginary of international development. In C. Mayo (Ed.), *Philosophy of Education* (pp. 353–361). Philosophy of Education Society.

Jakubiak, C. (2016). Ambiguous Aims: English-language Voluntourism as Development. *Journal of Language, Identity & Education, 15*(4), 245–258. https://doi.org/10.1080/15348458.2016.1195270

Judge, R. C. (2015). Emotion, volunteer-tourism and marginalised youth. In M. Blazek (Ed.), *Children's emotions in policy and practice* (pp. 157–173). Palgrave Macmillan.

Judge, R. C. (2016). Negotiating blackness: Young British volunteers' embodied performaces of race as they travel from Hackney to Zimbabwe. *YOUNG Editorial Group, 24*(3), 238–254. https://doi.org/10.1177/110330881562635

Kapoor, I. (2004). Hyper-self-reflexive development? Spivak on representing the third world 'Other'. *Third World Quarterly, 25*(4), 627–647. https://doi.org/10.1080/01436590410001678898

Kobayashi, A. (2003). GPC ten years on: Is self-reflexivity enough? *Gender, Place & Culture, 10*(4), 345–349. https://doi.org/10.1080/0966369032000153313

Kobayashi, A., & Peake, L. (2000). Racism out of place: Thoughts on whiteness and an antiracist geography in the new millennium. *Annals of the Association of American Geographers, 90*(2), 392–403. https://doi.org/10.1111/0004-5608.00202

Kobayashi, A., & Peake, L. (2007). Unnatural discourse. 'Race' and gender in geography. *Gender, Place & Culture,1*(2),225–243.https://doi.org/10.1080/09663699408721211

Kontogeorgopoulos, N. (2017). Finding oneself while discovering others: An existential perspective on volunteer tourism in Thailand. *Annals of Tourism Research, 65*, 1–12. https://doi.org/10/1016/j.annals.2017.04.006

Koot, S., & Fletcher, R. (2021). Donors on tour: Philanthrotourism in Africa. *Annals of Tourism Research, 89*. https://doi.org/10.1016/j.annals.2021.103216

Kothari, U. (2006a). An agenda for thinking about 'race' in development. *Progress in Development Studies, 6*(1), 9–23. https://doi.org/10.1191/14649963406ps124oa

Kothari, U. (2006b). Critiquing 'race' and racism in development discourse and practice. *Progress in Development Studies, 6*(1), 1–7. https://doi.org/10.1191/1464993406ps123ed

Lacey, G., Weiler, B., & Peel, V. (2016). Philanthropic tourism and ethics in charitable organisations: A case study in central Kenya. *Tourism Recreation Research, 41*(1), 16–25. https://doi.org/10.1080/02508281.2016.1108610

Lahman, M. K. E. (2008). Always Othered: Ethical research with children. *Journal of Early Childhood Research, 6*(3), 281–300. https://doi.org/10.1177/1476718X08094451

Martins, C. (2011). The dangers of the single story: Child-soldiers in literary fiction and film. *Childhood, 18*(4), 434–446. https://doi.org/10.1177/0907568211400102

Mason, C. L. (2016). Tinder and humanitarian hook-ups: The erotics of social media racism. *Feminist Media Studies, 16*(5), 822–837. https://doi.org/10.1080/14680777.2015.1137339

Matthews, S. (2017). Colonised minds? Post-development theory and the desirability of development in Africa. *Third World Quarterly, 38*(12), 2650–2663. https://doi.org/10.1080/01436597.2017.1279540

McEwan, C. (2001). Postcolonialism, feminism and development: Intersections and dilemmas. *Progress in Development Studies, 1*(2), 93–111.

Misi, S. (2016). Being white in post-2000 Zimbabwe: A reading of Eames' *Cry of the Go-Away Bird. Journal of Literary Studies, 32*(3), 98–108. https://doi.org/10.1080/02564718.2016.1235384

Mostafanezhad, M. (2013a). The geography of compassion in volunteer tourism. *Tourism Geographies, 15*(2), 318–337. https://doi.org/10.1080/14616688.2012.675579

Mostafanezhad, M. (2013b). The politics of aesthetics in volunteer tourism. *Annals of Tourism Research, 43*, 150–169.

Mostafanezhad, M. (2014a). Volunteer tourism and the popular humanitarian gaze. *Geoforum, 54*, 111–118. https://doi.org/10.1016/j.geoforum.2014.04.004

Mostafanezhad, M. (2014b). *Volunteer tourism: Popular humanitarianism in neoliberal times.* Ashgate Publishing Limited.

Mowforth, M., & Munt, I. (2009). *Tourism and sustainability: Development globalisation and new tourism in the Third World* (3rd ed.). Routledge.

Moyo, D. (2009). *Dead Aid: Why aid is not working and how there is a better way for Africa.* Farrar, Straus and Giroux.

Mudimbe, V. Y. (1988). *The invention of Africa: Gnosis, philosophy and the order of knowledge.* Indiana University Press.

Murris, K. (2013). The Epistemic Challenge of Hearing Child's Voice. *Studies in Philosophy and Education, 32*(3), 245–259. https://doi.org/10.1007/s11217-012-9349-9

Ndlovu-Gatsheni, S. J. (2013). *Empire, global coloniality and African subjectivity.* Ebook Central.

Nieuwenhuys, O. (2013). Theorizing childhood(s): Why we need postcolonial perspectives. *Childhood, 20*(1), 3–8. https://doi.org/10.1177/0907568212465534

Nyahunzvi, D. K. (2013). Come and make a real difference: Online marketing of the volunteering experience to Zimbabwe. *Tourism Management Perspectives*, 7, 83–88. https://doi.org/10.1016/j.tmp.2013.04.004

Pailey, R. N. (2019). De-centring the 'white gaze' of development. *Development and Change*, 51(3), 729–745. https://doi.org/10.1111/dech.12550

Peet, R. (2009). *Unholy trinity: The IMF, World Bank and WTO*. Zed Books Ltd.

Pilossof, R. (2009). The unbearable whiteness of being: Land, race and belonging in the memoirs of white Zimbabweans. *South African Historical Journal*, 61(3), 621–638.

Proyrungroj, R. (2017). Host-guest relationship in the context of volunteer tourism. *European Journal of Tourism Research*, 16, 177–200.

Quijano, A. (2010). Coloniality and modernity/rationality. In W. D. Mignolo & A. Escobar (Eds.), *Globalization and the decolonial option* (pp. 22–32). Routledge.

Rabbitts, F. (2012). Child sponsorships, ordinary ethics and the geographies of charity. *Geoforum*, 43(5), 926–936. https://doi.org/10.1016/j.geoforum.2012.04.006

Rideout, L. (2011). Representations of the 'Third World' in NGO advertising: Practicalities, colonial discourse and western understandings of development. *Journal of African Media Studies*, 3(2), 25–41. https://doi.org/10.1386/jams.3.1.25_1

Santoro, N. (2014). 'If I am going to teach about the world, I need to know the world': Developing Australian pre-service teachers' intercultural competence through international trips. *Race, Ethnicity and Education*, 17(3), 429–444. https://doi.org/10.1080/13613324.2013.832938

Scarth, A., & Novelli, M. (2019). Travel philanthropy and development. In R. Sharpley & D. Harrison (Eds.), *A research agenda for tourism and development* (pp. 88–109). Edward Elgar Publishing Limited.

Sensoy, O., & Marshall, E. (2010). Missionary girl power: Saving the 'Third World' one girl at a time. *Gender and Education*, 22(3), 295–311. https://doi.org/10.108009540250903289451

Sheridan, V., Landy, D., & Stout, V. (2019). The return of the 'Black Babies': How development education affects schoolchildren's attitudes to the majority world - the Trócaire 2012 Lenten campaign. *Race, Ethnicity and Education*, 22(6), 858–874. https://doi.org/10.1080/1361332.2017.1417829

Sinervo, A., & Hill, M. D. (2011). The visual economy of Andean childhood poverty: Interpreting postcards in Cusco, Peru. *The Journal of Latin American and Caribbean Anthropology*, 16(1), 114–142. https://doi.org/10.1111/j.1935-4940.2011.01127.x

Smithers, K., & Ailwood, J. (2022). School tourism in Southern Africa. In G. Noblit (Ed.), *Oxford Research Encyclopedia of Education*. Oxford University Press. https://doi.org/10.1093/acrefore/9780190264093.013.1811

Spivak, G. C. (1998). Can the subaltern speak? In C. Nelson & L. Grossberg (Eds.), *Marxism and the Interpretation of Culture* (pp. 271–313). University of Illinois Press.

Stainton, H. (2016). A segmented volunteer tourism industry. *Annals of Tourism Research*, 61, 256–258.

Sun, J., & Yanjun, X. (2020). The 'internal orientalism': New encounter in Tibet tourism. *Current Issues in Tourism*, 23(12), 1480–1492. https://doi.org/10.1080/13683500.2019.1658724

Suzuki, Y. (2017). *The nature of whiteness: Race, animals, and nation in Zimbabwe*. University of Washington Press.

Suzuki, Y. (2018). The good farmer: Morality, expertise, and articulations of whiteness in Zimbabwe. *Anthropological Forum*, 28(1), 74–88. https://doi.org/10.1080/00664677.2018.1429252

Tagwirei, C., & de Kock, L. (2015). Whitelier than white? Inversions of the racial gaze in white Zimbabwean writing. *African Identities*, 13(3), 184–198. https://doi.org/10.1080/14725843.2015.1023256

Taplin, J., Dredge, D., & Scherrer, P. (2014). Monitoring and evaluating volunteer tourism: A review and analytical framework. *Journal of Sustainable Tourism*, *22*(6), 874–897. https://doi.org/10.1080/09669582.2013.871022

Thompson, J., & Taheri, B. (2020). Capital deployment and exchange in volunteer tourism. *Annals of Tourism Research*, *81*. https://doi.org/10.1016/j.annals.2019.102848

Tomlinson, B. R. (2003). What was the Third World? *Journal of Contemporary History*, *38*(2), 307–321.

Wale, K., & Foster, D. (2007). Investing in discourses of poverty and development: How white wealthy South Africans mobilise meaning to maintain privilege. *South African Review of Sociology*, *38*(1), 45–69. https://doi.org/10.1080/21528586.2007.10419166

Wallerstein, I. (2000). *The essential Wallerstein*. The New Press.

Wearing, S., Mostafanezhad, M., Nha, N., Nguyen, T. H. T., & McDonald, M. (2018). 'Poor children on Tinder' and their Barbie Saviors: Towards a feminist political economy of volunteer tourism. *Leisure Studies*, *37*(5), 500–514. https://doi.org/10.1080/02614367.2018.1504979

Wearing, S., Young, T., & Everingham, P. (2017). Evaluating volunteer tourism: Has it made a difference? *Tourism Recreation Research*, *42*(4), 512–521. https://doi.org/10.1080/02508281.2017.1345470

Wijesinghe, S. N. R., Mura, P., & Bouchon, F. (2019). Tourism knowledge and neocolonialism - a systematic critical review of the literature. *Current Issues in Tourism*, *22*(11), 1263–1279. https://doi.org/10.1080/13683500.2017.1402871

Wright, H. (2014). Volunteer tourism and its (mis)perceptions: A comparative analysis of tourist/host perceptions. *Tourism and Hospitality Research*, *13*(4), 239–250. https://doi.org/10.1177/1467358414527984

Wydick, B., Glewwe, P., & Rutledge, L. (2013). Does International Child Sponsorship Work? A Six-Country Study of Impacts on Adult Life Outcomes. *Journal of Political Economy*, *121*(2), 393–436. https://doi.org/10.1086/670138

# 5 Building the set and producing 'Africa'

In the next three chapters, I now return to the metaphor that frames the empirical chapters of this book, that of the film of 'Africa'. In particular, I will use the key players in the making and producing of a film to explore the mechanisms which support and maintain the tourism in Matopo School. In the film industry, the role of the producer is to manage all elements of a film's production, until release. A producer's role might include overseeing set design, creative direction, and logistics. On the other hand, a film director creates a vision for the film and oversees guidance given to actors, communicates the vision to the crew and ensures all departments on the set function effectively. In line with the film metaphor, in this chapter I argue that the teachers enact the role of director, by ensuring the actors, the poor-but-happy children, are performing their roles correctly. The school founders and other tourism personnel, in contrast, play the role of producer. As such, they have a broader role of overseeing the vision the school hopes to portray. While the metaphor is somewhat reductionist in nature, it is a useful heuristic device that aims to provide a framing for understanding the complex nature of the implementation of tourism in the school.

## Set design: an image of Africa

Tourist perceptions of place can be formed through multiple representational media prior to the destination being experienced, with 'armchair travel' becoming more prevalent than ever before (Nelson, 2005, p. 155), particularly in light of global travel shutdowns. As part of these pre-formed perceptions of place, there are temporal and spatial associations with Minority World imaginaries of the African continent (Salazar, 2009; van Beek, 2003). The association, termed the 'image of Africa' (Mudimbe, 1988), is perpetuated in Minority World media and literature. For example, Staples explores an image of Africa in safari films from the 1920s to the early 2000s and argues there has been very little change over time, stating, 'the image-making practices of Minority World travellers in Africa remain embedded in nineteenth-century narratives and visual tropes of travel and exploration' (Staples, 2006, p. 393). I now explore an image of Africa from the perspective of tourism personnel and establish their

DOI: 10.4324/9781003407980-5

role in the movie metaphor: of film producers responsible for set design and the overall production of the movie.

In the metaphor and tourism imagery, the heterotopic image of Africa involves landscapes which include stark wilderness, wild animals, and a lack of technology (Brooks et al., 2011; Staples, 2006). Rice (2003) explores the construction of these landscapes, explaining that they are, 'impossible landscapes made up of images looted – as were so many museum artefacts – from real places' (p. 46). These images are heterotopic, they are symbols which have become disentangled from 'the worlds which gave birth to them' (Rice, 2003, p. 46) and as images they have been reconstructed and recombined in a way that defies the reality of their birth environments.

There is no doubt the tourism industry profits from continuously portraying a flattened image of Africa, by providing 'authentic' experiences of village and cultural aspects of 'Africa' (Garland, 2012; Hays, 2012; Saarinen, 2011). For example, in the traditional village at the Great Zimbabwe monument, Taru and Gukurume (2013) found that,

> The Karanga culture is packaged for tourists [*sic*] gaze, it has lost its authentic nature, and its value has been reduced to aesthetic whose relevance is derived from money paid by tourists. Furthermore, the culture is portrayed as unchanging and as backwards as the performers can imagine.
>
> (p. 69)

As part of the complexities of these representations of culture, in their study Taru and Gukurume (2013) identified that this type of tourism is praised by some tourism operators for continuing the culture by transmitting it to younger generations and to foreign tourists. For the Karanga culture, tourism is a double-edged sword as it simultaneously maintains and destroys culture (Taru & Gukurume, 2013). The frozen-in-time element of this type of tourism prevents a culture from moving forward and adapting, as all cultures do over time (Caton & Santos, 2009; Gonye & Moyo, 2018). The frozen-in-time element governs the image of Africa in this heterotopia, and all forms of tourism which promote 'authentic' culture must then conform to this homogenised timeless image.

The formation of a heterotopic space allows white Minority World tourists to sustain a fantasy, the fantasy being that they can find primacy and undisrupted nature in African countries. The fantasy includes a plethora of discourses which work to both enable and constrain the potential images that could be presented to the tourists. These heterotopic images are a hark back to the colonial period in which images of the African landscape were produced and distributed through travel writings and tales (Tucker, 2019; van der Duim et al., 2005). In turn, the image of barrenness informs the production and maintenance of the heterotopia, which is the landscape and setting of the movie in the metaphor which underpins this book.

Foucault (1967/1984) identified that 'the heterotopia begins to function at full capacity when men arrive at a sort of absolute break with their traditional time' (p. 6). Upon arrival at the school, the tourists enter a timeless time where they witness what are perceived to be 'relics' of a pre-colonial era, such as the children dancing. Further, this heterotopic image of the school allows for the tourist gaze to move from gazing on animals, to gazing on people (Crossley, 2012; Hays, 2012; Mandevani, 2014). The gaze dehumanises both children and adults in the school by relegating their roles to that of a one-dimensional figure, as explained through the metaphor.

In Zimbabwe, an image of untouched landscapes is associated with safari tours and other wildlife tourism. Safari tours are, in part, a colonial relic that allows the gaze of the tourist to transfer from gazing on animals while on safari, to gazing on people (Crossley, 2012). Many undertake packaged tours, in which the itinerary is curated carefully to include a combination of safaris, cultural tourism, entertainment, and food experiences. The all-inclusive nature of these tours allows a shift of the gaze from gazing on landscapes, such as Mosi-oa-Tunya (Victoria Falls), to gazing on animals in a game park and finally, gazing on people when local markets and villages are visited. The shift is a barely interrupted act, with little interrogation of the implications of doing so. It could be argued that the white tourist gaze colonises all that it gazes upon (Cocker, 2012; Taylor, 2015), which includes landscapes (Hays, 2012), people (Crossley, 2012), and food (Mkono, 2011).

The gaze is implicit in the itinerary of packaged tours, of which tour guides usually have very little control. I asked Greg, a white tour guide for Education Tours, how much capacity he had to change the way the Education Tours organised their tours. In his response, Greg describes the shift:

> Not really, because do not forget all of these people are sent this information prior. So, they do… they expect to see certain things… After the trip we get a report back and the guests are always asked what part of the trip is the best part. What I try and make sure I get is, 'I cannot pinpoint anything because every day had something to offer'. and I see that, and I am happy. But some people will say, 'the school touched me the most' or 'meeting the elephants' or 'the tour of the actual falls themselves'.

This transference of gaze is described by Crossley (2012, p. 247) as 'horrifying' and a consequence of the continued Othering and exotification of the Majority World (see also Aitchson, 2001; Simpson, 2004). Minority World imaginaries produce a romanticised image of Africa as a feminised, seductive, pristine landscape (van der Duim et al., 2005); with this imagery, and associated shift of gaze, common in the cinematic space (Staples, 2006). For example, the IMAX movie *Africa: The Serengeti* compares the migration of animals to the nomadic existence of the Maasai people (Salazar, 2009). In the context of New Zealand, Cocker (2012) refers to the Othering of Indigenous populations as something to be gazed upon, 'there is no question that the European coloniser viewed many practices of the indigenous population as picturesque or, more disturbingly, reflective of a lesser cultural group' (p. 27). The shift in gaze from animals, to people, is a result of supremacy of white ways of knowing and doing. Whiteness, as a pervasive discourse, ensures this shift is unquestioned and appears natural, the shift is also aided by the colonial histories of non-white people being classified as lesser. As an integral part of a heterotopic image of Africa, this shift dehumanises and sidelines people who become an object to be gazed upon.

The transference of gaze was not limited to the school grounds – Greg would often take his tour clients to the marketplace prior to a visit to the school. Whilst a market visit is not on the official tour agenda, Greg uses the market visit to provide further viewing of Otherness. Greg explains a visit to a local market,

> You know it is one of the things I sometimes do as well. Before we come out [to the school] we go to the market and I say to my guys, 'right let us buy 5kgs of beans or ten cabbages or 5kgs of onions and tomatoes'. And we donate that. So, they get something out of it, because they actually go to a market and they can haggle [**laughs**]. They can do all that side of it, and then they are giving it to the school as well, so that is... I try to encourage that more than actually giving money.

In this example, the tourists are using local people in the marketplace as an exotic 'Other' to experience their own visions of both authentic Zimbabwean life experiences and feel-good donations. The area that Greg brings the tourists from has at least three modern supermarkets; yet he takes the tourists to a place which is counter to the reality of most of the local residents, that is, those who shop at the supermarkets rather than the marketplace. Greg's decision making is placed within the dominant discursive framing of what 'Africa' should appear to be, and his choice is constrained by this framing.

In his choice to present a certain image of life in Matabeleland North, he is constrained by dominant discourses that depict 'Africa' as a heterotopia and a throwback to another time. Greg is also constrained by his position as a tour guide. As tours are pre-planned, tourists may arrive with a predetermined idea of what they will see – leaving very little discretion for tour guides. Foucault (1982/2002) identified the relationship between the exercise of power and the ability to govern others, 'to govern, in this sense, is to structure the possible field of action of others' (p. 341). In a sense, Greg is governed by the dominant discursive formations of what the tour *should* be, which structure the possible fields of action. Interestingly, he makes a choice which is counter to the tour itinerary – taking the tourists to the market – however, this decision is still framed through the expectations of how a market tour in 'Africa' might look. Further, Greg identified that each tour group provided a report on the tour, which Education Tours used to determine his proficiency as a guide. This added layer of surveillance may have meant Greg felt he had to present an image of Africa the tourists wished to see.

In the interview extract, it becomes apparent the reason Greg takes them to the marketplace is that they *get something out of it* and he constitutes the marketplace as quintessential Africa. The discourses of 'Africa' work to ensure that the marketplace is simultaneously a heterotopic place in which people work and do everyday tasks; but also becomes a place in which the tourists can gaze upon an ideal 'image of Africa', which includes a bustling marketplace, devoid of technology (Hetherington, 1997). The marketplace visit further develops the discourses associated with a heterotopic image of Africa, as it ignores the modern supermarkets in favour of a more traditional style marketplace. Images such as these are not false, but do not represent the *whole* picture of the culture/society/country they claim to portray. The heterotopic image, in this case, works to confirm stereotypes previously held about Zimbabwe – that of exotic marketplaces – and enables the tourist gaze to shift seamlessly from a gaze on landscapes to a gaze on people.

To extend the discussion further, this heterotopic image has a colonial connotation in which the Other are considered less civilised and less developed (McClintock, 1995). In colonial discourses, the Other were constituted as part of the landscape and as such under the control of white colonisers (Brandt & Josefsson, 2017; Suzuki, 2017). In this discourse of 'backwardness' there is a suggestion of unequal power relations in which the Other are subjectified and dominated, as they have always been considered 'less than'. In Greg's terming, the tourists 'get something' out of the marketplace, and by extension the local people. It is within this heterotopic image of Africa that colonial discourses are embedded and continue to govern the subjectivities of all in the scene. The extension of the local people as part of the wildlife/landscape renders them, at times, a voiceless participant in this form of tourism (Salazar, 2009). As Carter (2008) in their study of Cuba argues, the fetishization of the everyday lives of residents subjugates them into 'mere commodities of foreign imaginations and money' (p. 284).

In addition to the commodification of local people, haggling is a very specific part of the tourist experience. Hawkins (2010) explored haggling in the city of Tunis and found the stores with fixed prices were often unsuccessful due to the wish of tourists to have an exotic experience, that of haggling in a foreign marketplace, 'by successfully haggling for an item, the aspiring cosmopolitan has a double triumph, once over the local Tunisians, and once over other tourists who did not haggle successfully' (p. 15). Further, Kabwe-Segatti (2009) describes haggling as part of the 'game',

> The cultural background of the trader, the story he tells about the product, the contact he develops with his customers, the bargaining, the language he speaks are all part of the game and tend to transform the purchase into an experience that goes beyond the mere acquisition of goods.
>
> (p. 296)

Haggling represents an uneven exchange, where rather than allowing the local person to make some money from the transaction, tourists feel the need to haggle so they are not 'ripped off' by the perceived Other. The amount may make very little difference to the tourist, but for the lives of the marketplace sellers it could make a large difference. For example, monthly rent in Zimbabwe ranges on average from US$30–$120 (ZIMSTAT & World Bank, 2020). With many tourists haggling for a difference of US$5–10, this could have a significant impact on the ability for a marketplace seller to pay their rent. The image of the perceived Other, who is positioned, through discourses, to always be intent on scamming the white tourist (for an examination of this in Cuba see Simoni, 2014) extends into the school, this is discussed later in this chapter.

Following the reasoning of Hawkins (2010) who identifies that a successful haggle positions a tourist as an experienced and 'authentic traveller', the tourists are constrained by their subject positioning, that of the wise and experienced tourist who cannot be scammed. Foucault (1982/2002) identified power relations as,

> A total structure of actions brought to bear upon possible actions; it incites, it induces, it seduces, it makes easier or more difficult; in the extreme it constrains or forbids absolutely; it is nevertheless a way of acting upon an acting subject or acting subjects by virtue of their acting or being capable of action.
>
> (p. 341)

Thus, for tourists to constitute themselves as an experienced tourist, they may need to feel they have 'conquered' the local marketplace through successful haggling (Huxley, 2004). Power relations can constrain possible actions; in this instance, the tourists are constrained by the dominant image of the marketplace and their behaviour may be a reflection of the discourses which position the heterotopic marketplace as a place for tourists to be 'scammed'. The relational nature of power creates a relationship in which the tourists, who are more economically prosperous, are governed by the dominant discourses of how to be an 'experienced tourist'. Further, haggling provides another story or experience in the unfamiliar world of 'Africa' for the tourist to share when they return home, fulfilling the destination imagery of a place that is foreign to their home.

The tourism personnel in this setting are not ignorant to the associated discourses of men who will engage in crime and intimidation – and were quite reflective about the ways that their work contributed to this image. Greg, a white tour guide, encounters this image in his work with tourists,

> A lot of the time when I do guiding, what I try and do is: to tell the story of Africa, to give an idea of why Africa is the way it is. Because a lot of the time people, they get all the media. The media gets it wrong, there is this media bias that comes through. Everyone thinks of Africa as dangerous, full of gun toting guys with sunglasses on, very intimidating. That sort of thing. But there is a lot more to it than that.
>
> (Greg)

For Greg, who works for Education Tours, his aim is to counter the usual images of Africa as a problematic and dangerous continent. Greg sees the main concern for most tourists as the *gun toting guys with sunglasses* on. An unspoken component of this image is that the gun toting men are always dangerous (Martins, 2011; Salazar, 2009) and black in the Minority World imaginary (this includes child soldiers, see Bex & Craps, 2016). An interesting element of this heterotopia is it forms a dichotomy, on one hand there is the discourse of protection *of* the local people, but on the other hand the tourists need protection *from* the local people. The people in this image become simultaneously scheming to defraud the Minority World and they are also in need of saving from themselves (Durrheim et al., 2011; McClintock, 1995). They are reduced to be men in sunglasses, a trope which strikes fear and is also juxtaposed against the image of a smiling child. In some respects, tourism encourages the idea of the childlike native as a fix to the trope of men in sunglasses (Avraham & Ketter, 2017). This dominant discourse, of men as dangerous and children as needing protection, influences the conditions of possibility for thinking about the school and it governs the power relations between the white tourist and the marked Other. The dominant discourses constrain Greg in such a way that he actively works to resist the dominant image of 'gun toting men' – as he aims to create a tour that shows that *there is a lot more to it than that*. In this example, Greg is illustrating the ways in which tourism personnel, and the school teachers, are working to produce an image of Africa that is counter to the dominant and expected images. The ways Greg resists dominant discourses is counter to current understandings of the tourist gaze (e.g., Larsen, 2014; Urry & Larsen, 2011), as he is *not* conforming to the immediate stereotypes regarding 'Africa'. As I will argue further, current conceptualisations of the tourist gaze do not account for acts of resistance, such as the one shown by Greg.

In the movie metaphor, tourism personnel and the school founders act as producers of the film. In this role they are responsible for overseeing set design and the creative vision of the film. As detailed, Greg considered potential experiences that he could provide to tourists and aimed to meet their expectations of what an 'African' tour might be, and at times works to resist the dominant discourses. Thompson, the school founder and a black Zimbabwean tour guide, also considered the production of tourism from a guest-focussed perspective:

> You also have to look at the… [**pause**] my assumption is that not everybody [**pause**] will repeat that trip to Africa. For some people it is a once in a lifetime experience. That 10 minutes, or 15 minutes of dancing to some people is all they will remember in their minds when they go away. So, we have to look at it in the sense that, yes, it is taking time: it is probably asking the kids to do something that they would not just usually do when meeting someone. But you have to look at the guest side of things, and also think, these are the people who are helping us. Potential helpers,

some are already helping, what are they taking away? What is in their mind when they are back home? Something has to be a little memento that they can hold on to. And I think that dance, and the little kids flipping on their back when they, you know. I think it will stay. It has got that effect on some people.

(Thompson)

In part due to his extensive career in guiding, Thompson is acutely aware of the image that people seek when they book a trip to see a mythical, heterotopic, Africa. As part of the heterotopic nature of an image of Africa, time stands still within the school, as the school becomes a "simultaneously mythic and real contestation of the space in which we live" (Foucault, 1967/1984, p. 4). For Thompson, who is the school founder and an established tour guide, this was at the heart of the tourism at the school: the creation of a place is a real, but mythical experience.

Thompson highlights three points to consider: (a) this is not something the children would usually do, (b) the tourists are positioned as 'potential helpers' in all interactions and, (c) the tourism is based in the idea of a memento for the tourists to keep and take-away. For Thompson, the image of children dancing is the quintessential experience of culture and Africanness, a moment frozen in time for the tourist to keep, both in their memory and in photographs. When Thompson refers to the dancing as *it has got that effect on some people*, he is referring to the powerful, emotional response which can sometimes be associated with a heterotopic image of Africa. As part of this emotional response, the dancing children are integral in the vivid imagery constituted in this heterotopia; they represent a throwback to the colonial era images of 'wild savages' (dance can also be a challenge of colonial ideals, see Gonye & Moyo, 2018). In tourism, racial stereotyping continues to remain normalised and uninterrogated due, in part at least, to the colonial romantic discourses which circulate in the tourism industry (Tucker, 2019). These colonial discourses promote the idea of discovering an untouched culture and preserving this culture from change (Caton & Santos, 2009).

Amoamo and Thompson (2010) argue that postcolonialism is 'one of the most powerful means of re-examining the historical past and re-configuring our contemporary worldwide cultural concerns' (p. 36). Tourism has a relationship with postcolonialism as many postcolonial nations are tourism destinations with, 'many tourism practices and relationships based on past and present-day colonial structural relationships' (Amoamo & Thompson, 2010, p. 37). In the case of Matopo School, the image of the children dancing works to reproduce the colonial, Eurocentric image of the wild, primitive, untamed nature of the African landscape and people (Enevoldsen, 2003). In contrast to colonial perspectives of dance, rather than being considered as a negative and backward practice, the dancing becomes a spectacle and a relic to be observed. The spectacle is associated with a romanticism of the colonial periods and dancing is accepted only as a cultural display associated with primitive culture – rather than a legitimate dance form (Mowforth & Munt, 2009). Through the dancing, and flipping on their back, the children recreate and present the visceral experience of a heterotopia of Africa. A cultural dancing display is part of the setting and the image which the tourists are anticipating when they visit a school or tourism lodge in southern Africa.

Using Thompson's explanation as a guide, we can examine how an image of Africa can develop the guiding image of the heterotopia within the school. The school is at once a representation of its own culture and a representation of all African cultures. The performance by the school is potentially the only thing which will be remembered from the school tour and becomes part of the dominant discourses which circulate to govern what an 'African' school *should* look like. The image of the child flipping on their back ignores the infrastructure of the school, the use of technology within the school, and the existence of African cultures beyond *just* dancing and drumming. The lack of technology in this image enforces the illusion of a heterotopia that the continent of Africa is without technology and urban centres. Through the school tour, and the subsequent photography by tourists, an image of Africa as devoid of technology is circulated in the Minority World.

Thompson also asks, *what are they taking away?* and discusses the image he hopes they will retain. They are potentially *taking away* their own notions of saving and helping the African continent, all of which are invariably tied up in discourses of saving life and preserving cultures (discourses of development are explored further in Chapter Seven). The idea that they will retain *something* is part of a well-established discourse of owning Africa, in which they are 'taking a piece of Africa home with them' (Gardner, 2016, p. xii). By taking a piece of Africa home, the tourists are drawn into discourses of preserving and maintaining a heterotopic image of Africa which is furthered through the proliferation of photographs, often shared on social media profiles, and through travel stories (Toomey, 2017).

As already discussed, Greg, a white tour guide for Education Tours, aims to portray a different image of Africa – which is counter to the image of 'gun toting men in sunglasses'. Greg explains how he tries to encourage a different image of Africa,

A lot of the guiding I do, is I try and give people a background. A lot of that comes through the history of Africa, starting with the Cradle of Mankind, then the Bantu migrations. Then colonisation, which shaped who the people are. I am a storyteller, so when we come out to the village, I give people a lot more background before they come out. I explain the way the village works, how the leadership works, the cultural side of it, all the marriage and dowry. That sort of thing all comes into it. I think a lot of the times people are quite shocked, they go to Cape Town and it is very familiar to them. They come up here and they realise that they are in the third world. But to actually go to a village, even the people who have seen them on television it is always a wakeup call for them.

When Greg takes the tour groups to the village, he aims to provide a comprehensive background, which explains the historical and cultural factors that shape the present village structures and life. The extent to which he achieves this is unclear. In his interview Greg also described how the village is at odds with prosperity – and the modern world – and how it is used as something to have 'shock' value and to teach the children a lesson. Greg worked also as a private guide, offering bespoke tours of the local area. He details a common request from his private clients,

> You find a lot of families, particularly grandparents, want kids to realise that you know, their lifestyle, their school, their education, could be *very* different. And they say, 'well can you take these kids – because they are driving me wild – to a school and give them an appreciation.'

> (Greg)

In this extract, Greg describes how his private clients request for their children to be given an 'appreciation' of the life they have in the Minority World. The use of another's culture to discipline children originates from an extremely privileged position and reflects modern societies in which there is an Othering of cultures who are different as a means of exhibiting cultural capital (Hage, 1998; Judge, 2016). When the school and village are used as a means for Minority World children to understand their own privilege, there is an implicit assumption that the African village is contrary to life for children in the Minority World, and that the visit will give them an 'appreciation'. The comparison and 'appreciation' constitute local people as a marked 'Other', which also facilitates the transference of gaze described earlier in the chapter. By using the village as a prop, Greg reinscribes the discourses that postcolonial scholars aim to dismantle – namely that of modernity as a marker of progression towards 'development'. Modernity, and progressions of development, are Minority World constructs which privilege whiteness and white culture (Kothari, 2006a, 2006b; Mason, 2016).

Greg's interview was contradictory in places and represents the struggles of living in the postcolonial heterotopic present of Zimbabwe. Greg's contradictory use of the village exemplifies how these discourses are powerful and embedded around single stereotypical stories of Africa. Yapa (1996) identifies:

> Institutions and their discursive practices are the agents by which [subjects] are divided, classified, and subjected to normalization. Consider the categories of normal versus mad, normal versus criminal, normal versus pervert, normal versus poor, modern versus traditional, and developed versus underdeveloped. They are the products of specialized discourses that determine the shape, form and constitution of [subjects].

> (p. 712)

So powerful are the dominant discourses which create a heterotopic image of Africa, even as Greg *aims* to give the full story of Zimbabwean culture and history, he is inevitably drawn into the dominant discourse of Africa as Other. In this way, Greg is governed by the discourses which divide Africa and Other, which have a mutually constitutive function to govern both self and others. Greg's complex navigation of these dominant discourses, and the way he both disrupts and maintains an image of Africa, exemplifies a need for postcolonial theorising, in which there is a recognition of the complexities of navigating life in a postcolonial world.

The heterotopia which has been discussed in this section reinforces a transference of gaze through dominant discourses which position the local people as Other. This finding builds on previous work which outlines the dangers of tourism facilitating a transference of gaze (Crossley, 2012). The tourism personnel and school founders as 'producers' of the movie work to maintain and reproduce an image of Africa, in line with the expectations of the tourists. However, as shown above, there are resistances to these dominant discourses. I now turn to address the 'directors' – the teachers in the school and the conflicting roles they play in the production of the tourism.

## Managing conflicting roles of the school

In the school, the teachers and school founders worked behind the scenes to produce an image of the school that was acceptable for tourist consumption. Whilst the previous section outlined a heterotopic image of Africa, and the overarching image which influences this tourism, I now outline one tension in conducting tourism in a school: namely, balancing the disruptions of tourism with the benefits of tourism. As the directors of the 'movie set' the teachers tried to balance the disruptions to the school day but are ultimately bound by their obligations both as teachers and as directors of the 'movie'. I focus on this tension in depth. First, I outline the benefit to the school to explain why the teachers feel that they should work towards the ideal image of the African school. Second, I outline the effect of the disruptions which includes the ways the teachers manage the disruptions.

Imbued in the talk of teachers and school founders was the sense that Matopo School was different from others in the area. Thompson, one of the school founders, describes this as a 'shift':

> But I think our school has motivated staff, I hope we can do more motivation so that we get a lot more stuff done. Yeah. It is definitely a big shift, from neighbouring schools to bring their kids. Of course, the big inflow now also from town. They come to our school. So, I think we have already shaken the area a little bit.

Zimbabwe has long-term and ongoing issues with teacher retention and motivation due in part to inconsistent teacher salaries and school funding (Sadoma et al., 2015). Thompson references this when he says, 'I think we have already shaken the area a little bit'. The school can recruit qualified teachers as the funds from tourism allow them to offer salaries in US dollars rather than RTGS dollars – an unstable Zimbabwean currency. Due to the extreme economic turbulence I described in Chapter 2, a salary in the stable US currency is an appeal.

Zibusiso, a teacher with many years of teaching experience, details the difference between the school and others in the local area. Zibusiso identifies that tourism and funding make life easier,

> Where there are tourists, we have other benefits. Because if you look at the schools which are in the same place as ours [but without tourists], we find that things are difficult: paying fees and buying books and all the stuff. So, when we are in the school like this one, which has got tourists, some of the things become easy. Even when we attend workshops, some of those things in other schools become difficult because they have to finance them on their own. At times they struggle as they cannot *even* feed the children [through their feeding programs]. So here, where we have got the tourists, it is far better.

In the extract, Zibusiso refers to the school's feeding programme. The programme is unlike others in the area, as it includes children who are not up to date with their school fees. Zibusiso believes it is not only the teachers and children who benefit from the arrangement. Community-based tourism ventures are marketed as providing benefit to communities rather than individual beneficiaries (Moyo & Tichaawa, 2017). In this sense, the school is fulfilling part of its aim: to develop the local community.

Many of the teachers identified there were benefits for *all* when the school received tourists. Precious details,

> The tourism is beneficial. Like even with the Adventure Company now the money is getting to the villagers. This means wherever the tourists are, they leave groceries. I think the villagers like them. Everyone likes the visitors; everyone likes the tourists.

For Precious, community benefits were important. Precious was employed at the school despite not having the correct certificates to teach; Thompson and Alison employed her as part of the 'giveback' to the local community after she demonstrated teaching skills on a volunteer placement. She was undertaking the necessary training, with support from the school. Zibusiso further explains the benefit to the local community due to the land purchase for the school,

> It is a good benefit; it is really helping. Even for the ones not working, who are without their O-levels, high school certificates or qualifications. Because it is part of the agreement that when we give you the land you employ people from the community.

Zibusiso outlines the discursive construction of the school as one that ensures the local community and children are looked after, in line with the aims of community-based tourisms. The school was frequently constituted as needing to comply to the needs of tourists, due to the ongoing benefits of tourism for the school. A regime of truth includes, 'the types of relations which link together manifestations of truth with their procedures and the subjects who are their operators, witnesses or possibly objects' (Foucault, 2012/2014, p. 100). The idea that the tourism is beneficial, and needed, is part of the regime of truth established by developmentourism. This regime of truth provides an understanding as to why the teachers participate in the tourism – they feel it is the best option to fund the school.

Many teachers felt that visits by the tourists were not just beneficial due to physical resources but due to the recognition *for* the school:

> We really benefit from the visitors coming here. Right now, we have a shop over there. Sometimes they just pick one or two things. The money also helps us, even if they did not leave a round figure in the office, but they could have left something there, in the shop. Which is going to be a benefit to the school... just by visiting it gives the school a name. Because when somebody comes here, even if they do not leave anything, wherever they will go, they will always tell people they once saw Matopo School.
>
> (Nolly)

In response to a question about tourists who might not be donating large amounts, Nolly argues that by visiting the school the tourists may engage in future philanthropic venues. The role of the tourists will be explored in more depth in Chapter 7; however, it is important to note that the teachers were governed by discourses of hospitality and care in the way they both engaged with and spoke about the tourists. These dominant discourses of care are reflective of an act of hope, that one day the visit may turn into future donations or connections (Suzuki, 2017).

The school was visited frequently by tourists, with the visits in 2019 averaging to three per week, for every week of the school year. With every visit lasting between one to three hours, this had a significant level of disruption of the school. The teachers in their interviews, however, did not position the disruptions as necessarily negative. Bridget, a teacher, explained there were some disruptions that could not be planned for, and that these were not limited to tourist visits alone:

> Sometimes you may be called, maybe you did not know that there are visitors coming and they just want to come in at that particular time, or maybe someone has got sick, and maybe you are now asked to go and attend to her class or to his class, sort of things like that... Then you are called off the lesson and the time does not wait for you. It goes and that

subject is being interrupted. Then you are no longer going to be able to move onto the next subject now. Since you had already introduced the previous lesson, you will not leave it in the air, you have to finish it, so the next subject now is being disturbed.

Later in the interview, Bridget identified part of the problem being the crowded school curriculum, as she struggled to reach the dictated minimum number of hours for each subject. Other teachers also disclosed how they struggled to work their way through the curriculum. Bridget demonstrates the tension between conforming to dominant discourses about how the tourists should be treated and the pressure to conform to dominant discourses about how 'good teaching' should look. In her everyday professional life, Bridget tried to conform to two dominant, and potentially conflicting, discursive framings: firstly, that all educational disruptions should be managed by the classroom teacher and secondly, tourist visits were of economic and material value.

The school tour has different configurations depending on which tour company is visiting the school; however, in all tour configurations at least one teacher is taken from their classroom for a period of time, ranging between one and three hours. In the interviews, the teachers expressed minimal concern about the disturbances. In part, the lack of disturbances could be due to the organisation of a 'duty' roster – the teachers took turns in being the 'on duty' teacher and could therefore plan for the visits. Bridget details,

> I make sure the time I go out to attend to the visitors they [the children] have something to do. I do not just leave them, I utilise the time, I must make them be occupied at that particular time, so the time does not get wasted.

The duty roster was not a complete solution, however, as the children were often left unsupervised with work assigned. Work during this time consisted of textbook activities or self-guided reading tasks, with children expected to engage in self-regulation behaviours. These arrangements are not unusual for Zimbabwean schools and such work and expectations were consistent across the school for all children, ages 3–12. Only two classes, the two Early Childhood classes (ages 3–5) had formal supervision measures in place. The formal supervision, however, was limited to the other Early Childhood teacher checking on the class occasionally, or if that teacher was absent, children from Grade 5 supervising the class.

In previous discussions of both volunteer tourism and tourism in schools (see, for example, Ankomah & Larson, 2008; Baptista, 2012; Coghlan, 2015; Kontogeorgopoulos, 2017), however limited they currently are, there has been very little consideration for the ways that tourism disturbs school functioning. One notable exception is the study by Bargeman et al. (2018), who detail the impacts of volunteer tourism on one school in Ghana, identifying that volunteer tourists often hinder the progression of student as they interrupt the implementation of regular syllabus materials. Bargeman et al. also highlight the complexities of the situation, as the volunteer tourists in their study often provided resources and donations that supported the school. The teachers treated the disruptions of tourism as just another hurdle towards delivering all of their curriculum. I argue that, similar to previous studies of tourism in schools, the teachers perspective of the tourism shifted from something abnormal to just become part of the everyday functioning of the school (see also Smithers, 2023).

Other teachers also described the ways the dominant discourses governed teacher behaviour in the way they conducted the tour:

> Because sometimes if you like it, or you do not, it might drag. You will be saying 'I will take 30 minutes with [the tourists]'. but you will see yourself and you have taken about an hour or sometimes an hour and some minutes, trying, still trying to make sure the groups have had their entertainment, that they have gone round, they have seen the things properly and then you can say goodbye. It is time consuming. But some of the things you cannot rush. If someone is seeing the classroom and they are interested and they are seeing how the kids are working, they are trying to see the kids' books. It is one of those things. If you drag them away, it is no good also.
>
> (Ayanda)

Ayanda, one of the senior teachers, describes the tension she feels in conforming to the discourse of what an 'African' welcome should look like. She identifies how some tourists are interested in seeing the schoolwork children are doing and in witnessing how a Zimbabwean classroom functions. The tour may take longer as the tourists sometimes linger in the classrooms and as a result, the teacher on duty is out of their own classroom for a greater length in time. By conforming to the need for the tourists to have a sustained interaction in the classroom, the school must endure sustained interruptions to the classrooms; however, Ayanda sees this as a necessary burden, saying *it is one of those things*. As part of the production of developmentourism, in which there is an emphasis on commodification, the teachers recognise that tourists are customers who may expect a level of service or hospitality throughout the tour (Baptista, 2011, 2012). Whilst the school functions in the way of a school with timetables, bells, and break times, the ways of doing in the school are also shaped by the need to conform to tourist expectations – which creates an unusual dynamic within the school that has become part of the everyday functioning. Ayanda, in her role as a movie director, conforms to the dominant discourses which constitute tourists as the priority for the school.

Some teachers did not ascribe to the dominant discursive regime of tourism always being for benefit. One teacher questioned the need to involve children in the process:

> What I am saying would be, we are so desperate for the funds that they are giving, but maybe in the near future people will realise that, 'oh what we have been doing to other people's kids? It is not nice', and then we will stop the travellers taking photos of the kids.
>
> (Astrid)

Astrid identifies the tension between the two dominant discursive positions, that the school is *desperate for funds* and therefore must conform to the expectations of the tourists, but also that she is a teacher with responsibilities of care for her students. At the basis of all teacher interviews was a sense of obligation to donors, due to the funds they provided to the school. The reliance on benefits of the tourism, namely monetary, is reflective of the economic turbulence in Zimbabwe which contributes to a dependence on donor aid. Further, previous studies (Arnett, 2016; Hernandez, 2018) have established, 'the protocol for donor-beneficiary encounters is a familiar one' (Suzuki, 2017, p. 137). The dominant discourses frame the benefits of tourism as integral to the functioning of the school, which forms a regime of truth through which the teachers and children are governed. A regime of truth does not always represent something which is 'true' – rather it is produced *as* true by a regime of power (Foucault, 2012/2014). Astrid was one of the few teachers who willingly critiqued the behaviour of tourists in her interview. She also invokes a child protection concern which has been raised in other discussions of photography in tourism (Arnett, 2016; Hernandez, 2018; van Beek, 2003).

Thompson, a school founder, expressed his own caution about the level of comfort and interaction the children had with the tourists. He described the measures in place to try and protect the children and identified how things would look different as the children aged,

> I have a feeling that as the kids get older, get into secondary school, there is a lot of supervision to be done. But we would have to have a lot of control. Because with teenagers, you know they start writing to their… to the guests. We just have to be vigilant and be ready for, when things start changing. Especially with the secondary school students. I think, we should give the guests the opportunity to go to the secondary school but there is a lot of supervision required.
>
> (Thompson)

Thompson identified how the current nature of the tours may not be appropriate for a secondary school setting, recognising there is a risk to the children in the school. Thompson, like the teachers at the school, is governed by the dominant discursive framing of how tourists should be treated. It could be argued that the teachers and school founders were unwilling to engage in critiques of tourism, as this would involve pushing back against dominant discourses which privilege the interventions by tourists. Subverting dominant discourses involves a level of risk and for the teachers it could mean a loss of job. For Thompson, subverting the dominant discourses, which privilege service and hospitality, could mean loss of income for the school or lack of business for his own safari company.

The teachers seem to accept the tourism as beneficial as the tourism is framed through dominant discourses about how donor-beneficiary encounters *should* look: namely that the beneficiary should be accepting of the donation no matter how useful the donation is to their life (Chouliaraki, 2010; Hernandez, 2018). The discourses legitimise a regime of truth: tourists are always beneficial as they bring resources and monetary donations and if they do not bring physical donations, they give the school a name which may facilitate future donations (this perspective is explored in the next chapter). The regime of truth enables certain behaviours and privileges some voices over others. In this case, the school identifies the tourists as being a key source of income and their voice gains more authority than those of the teachers.

To return to the metaphor, the teachers and school founders are constrained by their positioning as producers and directors of the 'movie'. In some ways, they have limited control of the creative direction of the set design, as it has been predetermined through the proliferation of a heterotopic image of Africa. In another sense, their actions are framed through dominant discourses of how a tourist should be treated when they visit the school. Like a school depicted on a movie set, it does not matter to the tourists if the school is functioning how a school should, it only needs to appear to be a functioning school. To expand on this notion, the school that the tourists see is not necessarily the school that the children see in their everyday lives. The children and teachers act in particular ways during tourist visits, in ways that are not always representative of the day-to-day functioning of the school – although the disruptions of the tourism have become embedded into the schools routines and self-governance. Building on the idea of the teachers and founders being producers and directors of an image of Africa, I now explore the deliberate use of children to conduct the school tour.

### Use of children because teachers tell lies

There was an interesting contradiction in the research data surrounding the relationship between the tourists and the teachers. The teachers I spoke with voiced the idea that there was a mistrust of the teachers at the school by tourists. In various ways the teachers identified the ways they governed children's behaviour, and their own, to enable the tourists to see how their money was, or would be, spent. At the heart of these endeavours was an aim to increase the overall perceived trustworthiness of the school, in the hopes that this would bring more donations. The tourist gaze has previously been conceptualised as one-way with tourists gazing upon the exotic other (Larsen, 2014; Perkins & Thorns, 2001; Urry, 1990; Urry & Larsen, 2011). I now provide a counter to this one-way perspective of tourist gaze. I argue the tourist gaze governs the teachers and children in the school; but on the other hand, the school community return a gaze of their own – which shapes their assumptions regarding tourists.

During the fieldwork, Astrid recounted a story to me about a tour group who had visited the school and paid for their purchases in the school shop with US$1 notes. Astrid laughed as she detailed how the tourist had been very careful to hide the contents of their purse from Astrid, as they counted out the money to make the purchase. In the act of hiding the contents of the purse, the tourist had spilt the money. Astrid said to me,

> The lady she was spilling them everywhere. She had only opened her purse the tiniest bit. She did not want to show me what was inside

Throughout the fieldwork, the teachers detailed how they were aware of the image that the tourists had of them: that they were untrustworthy. Power relations are a network of relations that "organise and define objects" (Phillips, 2002, p. 335), which in turn enable and constrain ways of acting and speaking. The perception of local people as scheming to defraud tourists is reflective of a subject positioning which is powered through discourses associated with a heterotopic image of Africa. To borrow from Phillips (2002) terminology, teachers recognise the way they are located within the power relations and the discourses that work to define them and in return, engage in behaviours to maximise funding for the school. As such, the teachers become *the directors* on the movie set as they work to create their heterotopia and an image of Africa, whilst using the children in the school to convey the message of a desperate need for funding (a well established technique in NGO advertising see Rideout, 2011).

A heterotopia built on the image of Africa, and the associated discourses, creates both enabling and constraining factors on the ways of communicating in the school. In this example given above, Astrid understands the tourists think the teachers are dishonest, as demonstrated through the act of hiding the notes in their wallet. Tourists may think the teachers are deceitful as Minority World media privileges the dominant discourses which position all people who live in African countries as a homogenous group who wish to defraud or thieve from innocent tourists (Schwarz & Richey, 2019). The dominant discourse of dishonest local people is a throwback to colonial tropes regarding the exotic 'Other', in which the Other was perceived as deviant, mysterious, and childlike (Aitchson, 2001; Caton & Santos, 2009). The teachers actively work to avoid the tourists perceiving them as untrustworthy.

In the interview with Astrid, following on from the encounter outlined above, we had the following exchange:

*Interviewer:*    Do you think that, sometimes… they think they are going to get scammed or something if they talk to the teachers?

*Astrid:*    I think so.

*Interviewer:*    Yeah, I do remember you telling me a story about someone who came, and they had their wallet, they were trying to get money out and then it just spilt everywhere because they were trying to be all secretive about it… **[both laugh]**

*Astrid:*    That happens, they prefer the kids rather than the teachers.

There is a dominant image of adults in the Majority World as thieving and untrustworthy. In tandem with the image of adults, there is a dominant discourse that positions children in the Majority World as needing care and protection. In these images, the protector is usually an adult from the Minority World, rather than an adult in the child's own country. Foundations such as 'Save the Children' work to protect children worldwide but in doing so further this discourse of children needing protection from unsafe adults in their home country (Pailey, 2019; Rideout, 2011).

Zibusiso also outlined a perceived mistrust of how the teachers would be using the donations:

And most of the things when they are using the things, it must be children, children, children. No mention of the teacher. Actually, the teacher is like a stumbling block, they think maybe teachers will steal things from the children. Which is I think a negative attitude and it is really a setback, because if you entrust with children, entrust with everything. But it is like small things, when they come back, they want to look at everything to say, is this still there?

(Zibusiso)

Zibusiso then detailed how it was unfair for such as an expectation to exist, when children are rough with items which might mean items are destroyed unintentionally. The teachers perceive that there is a mistrust on the part of the tourists and that if items are destroyed by children, the teachers will be asked what happened to it. In Zibusiso's understanding, the tourists hold a deficit view of teachers which constitutes the teachers as untrustworthy. Foucault (1975/1991) argued:

We must cease once and for all to describe the effects of power in negative terms: it 'excludes', it 'represses', it 'censors', it 'abstracts', it 'masks', it 'conceals'. In fact, power produces; it produces reality; it produces domains of objects and rituals of truth.

(p. 194)

The dominant discourses, of untrustworthy adults and needy children, work to produce a particular reality which is framed through a heterotopic image of Africa. In this way, thinking about the Majority World, tourists may be both enabled and constrained by the discourses that circulate. The dominant discourses, which are often played into by the school, position both the tourists and the children in particular ways – which leads both the tourists and school teachers to govern their behaviour regarding conduct on the tour. In this way, the discourses are part of producing a reality that all involved must conform to. The school recognises this and the need for children, rather than adults, to play the role of tour guide and messenger, as a strategy to increase funding by playing on the existing stereotypes and discursive constructions about the Majority World.

The perceived mistrust of teachers by tourists was integral to the management of the school tour. Astrid details how the teachers' role works in the school tour:

> Like the teacher will be there but they will not be of help. He will be, like following the group, he will not be saying much, because the visitors they prefer talking to the kids than the teacher. Because they think the teacher will be telling them lies and all [laughs]. So, they want to get it from the learners, rather than the teachers.
>
> (Astrid)

Astrid identifies that the school uses the children as a means to support the school's message of needing support. The children used as guides were frequently displayed in the children's drawings, which showed children walking hand in hand with tourists. For example, drawings (Figures 5.1 and 5.2) by Tawanda and Peace (age 9) highlight this.

In the drawing by Peace there is a depiction of a school bus on the left with two children and two tourists on the right. The dialogue between one child and a tourist is the tourist asking, 'What is your name?' and the child replying, 'My name is…' (name removed to protect child identity). In the picture by Tawanda, there is a kitchen hut drawn in the middle, with two children sitting at a table

*Figure 5.1* Drawing by Peace depicting a school bus with tourists and children holding hands.

*Figure 5.2* Drawing by Tawanda depicting a brick building with a tourist and child holding hands.

to the right. There are two figures to the left, one larger figure (a tourist) and one smaller figure, they are holding hands. A tourist asks, 'who is your friend?' and the child responds (name removed to protect identity). There is a caption that identifies the location as the kitchen.

The use of child guides could be to capitalise on the discourses of innocence which are associated with Minority World constructions of childhood. Children are used in marketing of volunteer tourism and orphanage tourism to invite individuals to 'help children, who because they are innocent, cannot be held accountable for their own poverty' (Carpenter, 2015, p. 20). A positioning, of children as innocent and trustworthy, is part of the dominant discourses of children in advertising in which they are either pictured on the dichotomy of poor-but-happy or fly-blown poverty-stricken children (Caton & Santos, 2009; Crossley, 2012; Etang et al., 2012; Toomey, 2017). Child-soldiers are contrary to this image – as they are perceived dually: both as knowing the ramifications of their actions and as being duped into participating in war due to their innocent nature. The contradictory images of children in the Majority World are explored in the next chapter – of interest for this chapter is the deliberate choice of the children as part of the tour *by* the teachers. The choice to use child guides is reflective of the ways in which the teachers gaze back at the tourists. Rather than subverting to the dominance of the tourists – as would be expected through the dominant conceptualisations of tourist gaze in tourism literature (Larsen, 2014; Urry & Larsen, 2011) – the teachers at the school are involved in substantial reflection regarding the expectations of the tourists.

To illustrate this point, Nolly describes the way the school ensures that the tourists believe what the teachers are telling them:

> The [children] only do the beadwork when the visitors are around. That keeps them doing something, other than to be idle, and one of those reasons is that we wanted them to portray what they do. Because we cannot say, 'the children are doing this – it is the children who do this' – whilst they have never seen them doing that very thing. If they find them doing that same thing, at least they know that it is done by the children.
>
> (Nolly)

Nolly identifies that if they do not see it, the visitors may not believe the children are the ones who make the beads. The pervasive nature of the tourist gaze acts as a normalising technology of power. The school *feels* that they must have a level of performativity to the beading. It is not enough for the children to have done the beading; it needs to be proven through the visible act of the children beading. The beading was depicted in a picture (Figure 5.3) by Gareth.

*Figure 5.3* Drawing by Gareth depicting the school jewellery shop.

In this drawing, two tourists are depicted in the school 'shop' and are commenting on the jewellery made by the children. They say, 'oh! Good' and 'oh! nice beads'.

It could be argued that the tourist gaze, in this case, works to govern the behaviour of the children during the school tour. Teachers are hyper-aware of the need to conform to the needs and wants of the tourists, so they engage in behaviour which encourages a level of performativity. Urry and Larsen (2011) position this level of performativity as a way of engaging with tourists that, 'broadly signals the power of the tourist gaze in orchestrating tourist service encounters' (p. 72). I would argue that, based on the examples provided in this chapter, whilst the tourist gaze *is* powerful, in this case it appears to be not as far-reaching as conceptualised by Urry and Larsen. The power relations in the school are complex, with the school community returning a gaze of their own.

Zibusiso detailed how the teachers worked to ensure donors could see their money was spent in the intended manner. The teachers would usually take pictures of the purchase:

We have got someone who is donating money for the shoes. He is still waiting for the shoes; I think that person is donating about $100 per month. We have not been able to purchase those shoes. But he is waiting for the photos to see children putting on those shoes. Because when he started donating, the economy – things started skyrocketing, prices skyrocketing, things started running short from the shelves – we have not been able to buy the shoes because of these reasons.

Despite having a reason for not purchasing the shoes, that is, the inability to purchase them due to the economy, there is still pressure from the donor for the shoes to be bought. Zibusiso later detailed how the shoes would need to be purchased in Zambia and brought across the border for the school to purchase them at a reasonable price. In addition to the time required to visit Zambia, there are added costs as Zimbabwean citizens can only bring a certain amount of goods per month before needing to pay tax on the items brought into the country.

Zibusiso's story problematises the notion that any donation is a useful donation, a well-established discourse of Development (Arnett, 2016; Baptista, 2011; Sun & Yanjun, 2020). Requesting evidence of children wearing shoes is a particular form of surveillance that constitutes donations and donor recipients in a certain manner. As part of the surveillance, the donor would like evidence that the donation has been purposefully used and appreciated by the recipient, which constitutes the relationship as unequal (Chouliaraki, 2010; Hernandez, 2018; Thompson, 2022). The school is placed within unequal power relations, in which the school must conform to the dominant discourses of what a recipient will do – send pictures of the items being used by a grateful child – even if this task is difficult due to the failing economy. Although they are placed within unequal power relations, the school is not trapped within the power relation, as

Foucault (1980/2002) argued, 'those who are enmeshed, involved, in these power relations, can, in their actions, their resistance, their rebellion, escape them, transform them, in a word, cease being submissive' (p. 294). In the unequal relation, the school does practice a resistance; they have not purchased the shoes despite the donor's insistence. There are complexities to monetary donations, particularly if the recipient is required to show proof the funds have been well-spent (the complexities of gift-giving are explored in depth in Chapter 7).

Later in the interview, Zibusiso further detailed how the school had sent pictures of a child in a uniform to a donor,

> I have got one [donor] in my [phone] contacts who bought uniforms for a child and they actually said, 'send me photos'. I actually took the photo, I sent it, but I have not got a response. So actually, they like seeing what they have donated.

> (Zibusiso)

During the fieldwork there were several stories like this one, where the relationship did not extend beyond the initial donor-recipient obligations. This demonstrates the fleeting nature of some of these interactions. Like the audience of a movie, the tourists visit and are immersed for a short period; however, the experience is fleeting and may include no follow-up or response. In the volunteer tourism literature, there have been debates regarding the impact on people who visit schools or engage in these types of tourism. There are scholars who argue that the 'life changing experiences' of volunteer tourists are reflected in their day-to-day lives upon returning to their home country (Brown, 2005; McGehee & Andereck, 2008). In contrast, some argue that volunteer tourists have very little engagement beyond their tourism visit and the 'life changing experiences' they have are fleeting at best (Bailey & Russell, 2012; Bauer, 2017; Crossley, 2012). In the case of Matopo School, the stories of Zibusiso suggest that the interactions between the teachers and tourists are fleeting. Despite being positioned as an 'authentic' experience that allows the tourists to have a connection with children and staff, there are questions to ask about whether this interaction goes beyond a brief encounter. A counter reading of the situation is to consider whether there is a need for the encounter to be lasting at all. Beyond the immediate surveillance of determining money is well spent, is there a need for donors to continue their relationship?

I have outlined the ways in which the school perceives a mistrust of their ability to use donations in a worthwhile manner. Arguably, the perceived mistrust governs their behaviour in terms of how they conduct the tour and the experiences they provide to the tourists. For example, the children are used to demonstrate the purposeful use of donations. The teachers, as directors of the movie, are responsible for the continuation of donations and funding. In the making of an image of Africa, the teachers place the children at the forefront of the tour, due to the Minority World tropes about untrustworthy adults in the Majority World.

## Summary

There are dominant discourses, entangled in a heterotopic image of Africa, which work to create tropes about what the tourists might see on an 'African' school tour. These tropes govern the behaviour of all in the setting. The Othering of Matopo School is enabled by a heterotopia based in the image of Africa that forms the basis of all interactions in the school.

There is a tension between being a school and a tourism destination. The teachers are in an awkward professional position in which they negotiate between conforming to discourses of what a 'good' teacher looks like and discourses of what a 'good' African welcome looks like. They navigate these roles within an understanding of what the tourists are hoping to see when they visit the school.

I have also outlined the way the teachers feel mistrusted by tourists. In order to maximise funding, the school teachers use the children in the school to portray the needs of the school. For the teachers at the school, the tourists are considered a positive influence, but they are also perceived to have a distrust of the teachers. Aitchson (2001) argues that in tourism, power is located solely in the white capitalist/imperialist hands. However, I argue relationships between tourist/host community are complex, and the school, as the producers/directors of the movie, actively construct an image of themselves so as to maximise donations to the school. By understanding the relationship as complex, in which power is relational rather than top-down, I have added a complexity to our understanding of typical host/guest relationships.

The mistrust of the teachers will be analysed further in the next chapter, as it is part of the ways the teachers govern the children, by constituting them as the poor-but-happy children. The next chapter explores the dominant discourses which constitute tourists as friends and positive influences on the school, but also as people who do not trust the teachers at the school. On one hand, there is a high level of trust for the tourists as they are welcomed into the school, but this trust is not returned by the tourists.

## References

Aitchson, C. (2001). Theorizing Other discourses of tourism, gender and culture: Can the subaltern speak (in Tourism)? *Tourist Studies*, *1*(2), 133–147.

Amoamo, M., & Thompson, A. (2010). re(Imaging) Māori touism: Representation and cultural hybridity in postcolonial New Zealand. *Tourist Studies*, *10*(1), 35–55.

Ankomah, P. K., & Larson, T. (2008). Creativity in cultural tourism: The case for rural development in sub-saharan Africa. *Tourism Review International*, *12*(3), 171–186.

Arnett, J. (2016). Taking pictures: The economy of affect and postcolonial performativity in NoViolet Bulawayo's We Need New Names. *Ariel: A Review of International English Literature*, *47*(3), 149–173.

Avraham, E., & Ketter, E. (2017). Destination image repair while combatting crises: Tourism marketing in Africa. *Tourism Geographies*, *19*(5), 780–800. https://doi.org/10.1080/14616688.2017.1357140

Bailey, A. W., & Russell, K. C. (2012). Volunteer Tourism: Powerful Programs or Predisposed Participants? *Journal of Hospitality and Tourism Management*, *19*(1), 123–132. https://doi.org/10.1017/jht.2012.14

Baptista, J. A. (2011). The tourists of developmentourism– representations 'from below'. *Current Issues in Tourism, 14*(7), 651–667. https://doi.org/10.1080/13683500.2010.540314

Baptista, J. A. (2012). The virtuous tourist: Consumption, development, and nongovernmental governance in a Mozambican village. *American Anthropologist, 114*(4), 639–651. https://doi.org/10.1111/j.1458-1433.2012.01515.x

Bargeman, B., Richards, G., & Govers, E. (2018). Volunteer tourism impacts in Ghana: A practice approach. *Current Issues in Tourism, 21*(13), 1486–1501. https://doi.org/1 0.1080/13683500.2015.1137277

Bauer, I. (2017). More harm than good? The questionable ethics of medical volunteering and international student placements. *Trop Dis Travel Med Vaccines, 3*(5), 5. https://doi.org/10.1186/s40794-017-0048-y

Bex, S., & Craps, S. (2016). Humanitarianism, testimony, and the white savior industrial complex: What is what versus Kony 2012. *Cultural Critique, 92*, 32–56.

Brandt, F., & Josefsson, J. (2017). Sexuality and power on South African game farms; reflections on positionality and emotions in ethnographic research. *Emotion, Space and Society, 23*, 26–32. https://doi.org/10.1016/j.emospa.2017.02.004

Brooks, S., Spierenburg, M., van Brakel, L., Kolk, A., & Lukhozi, K. B. (2011). Creating a commodified wilderness: Tourism, private game garming, and 'third nature' landscapes in Kwazulu-Natal. *Tijdschrift voor Economische en Sociale Geografie, 102*(3), 260–274. https://doi.org/10.1111/j.1467-9663.2011.00662.x

Brown, S. (2005). Travelling with a Purpose: Understanding the Motives and Benefits of Volunteer Vacationers. *Current Issues in Tourism, 8*(6), 479–496. https://doi.org/10.1080/13683500508668232

Carpenter, K. (2015). Childhood studies and orphanage tourism in Cambodia. *Annals of Tourism Research, 55*, 15–27. https://doi.org/10.1016/j.annals.2015.08.010

Carter, T. F. (2008). Of spectacular phantasmal desire: Tourism and the Cuban state's complicity in the commodification of its citizens. *Leisure Studies, 27*(3), 241–257. https://doi.org/10.1080/02614360802018806

Caton, K., & Santos, C. A. (2009). Images of the Other. *Journal of Travel Research, 48*(2), 191–204. https://doi.org/10.1177/0047287509332309

Chouliaraki, L. (2010). Post-humanitarianism Humanitarian communication beyond a politics of pity. *International Journal of Cultural Studies, 13*(2), 107–126. https://doi.org/10.1177/1367877909356720

Cocker, A. R. (2012). Picturing colonial New Zealand: Charles Spencer's images of the 'eighth wonder of the world', the pink and white terraces. *The International Journal of the Image, 2*(4), 22–32.

Coghlan, A. (2015). Prosocial behaviour in volunteer tourism. *Annals of Tourism Research, 55*(1), 46–60. https://doi.org/10.1016/j.annals.2015.08.002

Crossley, É. (2012). Poor but Happy: Volunteer Tourists' Encounters with Poverty. *Tourism Geographies, 14*(2), 235–253. https://doi.org/10.1080/14616688.2011.611165

Durrheim, K., Mtose, X., & Brown, L. (2011). *Race trouble: Race, identity and inequality in post-apartheid South Africa.* Lexington Books.

Enevoldsen, K. (2003). See no evil, hear no evil: An "Outsider's" encounter with cultural tourism in South Africa. *Cultural Studies ↔ Critical Methodologies, 3*(4), 486–502. https://doi.org/10.1177/1532708603253578

Etang, A., Fielding, D., & Knowles, S. (2012). Giving to Africa and perceptions of poverty. *Journal of Economic Psychology, 33*, 819–832. https://doi.org/10/1016/j.joep.2012.03.003

Foucault, M. (1967/1984). Des espace autres [Of other spaces: Utopias and heterotopias]. *Diacritics, 16*(1), 22–27.

Foucault, M. (1975/1991). *Surveiller et punir: Naissance de la prison* [Discipline and Punish: The Birth of the Prison] (A. Sheridan, Trans.). Penguin Books Australia Ltd.

Foucault, M. (1980/2002). Interview with Michel Foucault (conducted by D. Trombadori) (Robert Huxley and others, Trans.). In J. D. Faubion (Ed.), *Dits et ecrits 1954–1984 [Essential works of Foucault 1954–1984]* (Vol. Vol. 3: Power, pp. 223–238). Penguin Books Ltd. (Round table of 20 May 1978).

Foucault, M. (1982/2002). The subject and power (Robert Hurley and others, Trans.). In J. D. Faubion (Ed.), *Dits et ecrits 1954–1984 [Essential works of Foucault 1954–1984]* (Vol. 3: Power, pp. 326–348). Penguin Books Ltd.

Foucault, M. (2012/2014). *On the Government of the living: Lectures at the Collège de France 1979–1980* (G. Burchell, Trans.; F. Ewald & A. Fontana, Eds.). Palgrave Macmillan.

Gardner, B. (2016). *Selling the Serengeti: The cultural politics of safari tourism*. The University of Georgia Press.

Garland, E. (2012). How should anthropologists be thinking about volunteer tourism? *Practicing Anthropology, 34*(3), 5–9.

Gonye, J., & Moyo, N. (2018). African dance as an epistemic insurrection in Postcolonial Zimbabwean Arts education curriculum. In A. M. Kraehe, R. Gaztambide-Fernández, & B. S. Carpenter II (Eds.), *The Palrgrave Handbook of race and the arts in education*. Palgrave Macmillan. https://doi.org/10.1007/978-3-319-65256-6_9

Hage, G. (1998). *White nation: Fantasies of white supremacy in a multicultural society*. Pluto Press Australia.

Hawkins, S. (2010). Cosmopolitan hagglers or haggling locals? Salesmen, tourists, and cosmopolitan discourses in Tunis. *City & Society, 22*(1), 1–24. https://doi.org/10.1111/j.1548-744X.2009.01027.x

Hays, C. M. (2012). Placing nature(s) on Safari. *Tourist Studies, 12*(3), 250–267. https://doi.org/10.1177/1468797612461086

Hernandez, K. (2018). Can the subaltern be seen? Photographic colonialism in service learning. *Qualitative Research Journal, 18*(2), 190–197. https://doi.org/10.1108/QRJ-D-17-00051

Hetherington, K. (1997). *The badlands of modernity: Heterotopia and social ordering*. Routledge.

Huxley, L. (2004). Western backpackers and the global experience: An exploration of young people's interaction with local cultures. *Tourism, Culture & Communication, 5*, 37–44.

Judge, R. C. (2016). Negotiating blackness: Young British volunteers' embodied performaces of race as they travel from Hackney to Zimbabwe. *YOUNG Editorial Group, 24*(3), 238–254. https://doi.org/10.1177/110330881562635

Kabwe-Segatti, A. W. (2009). "We offer the whole of Africa here!". African curio traders and the marketing of a global African image in post-apartheid South African cities. *Cahiers détudes Africaines, 49*(193–194), 285–308. https://doi.org/10.4000/etudesafricaines.18718

Kontogeorgopoulos, N. (2017). Finding oneself while discovering others: An existential perspective on volunteer tourism in Thailand. *Annals of Tourism Research, 65*, 1–12. https://doi.org/10/1016/j.annals.2017.04.006

Kothari, U. (2006a). An agenda for thinking about 'race' in development. *Progress in Development Studies, 6*(1), 9–23. https://doi.org/10.1191/14649963406ps124oa

Kothari, U. (2006b). Critiquing 'race' and racism in development discourse and practice. *Progress in Development Studies, 6*(1), 1–7. https://doi.org/10.1191/1464993406ps123ed

Larsen, J. (2014). The Tourist Gaze 1.0, 2.0, and 3.0. In A. A. Lew, C. M. Hall, & A. M. Williams (Eds.), *The Wiley Blackwell companion to tourism* (pp. 304–313). John Wiley & Sons Ltd.

Mandevani, K. M. (2014). Kushanya Mumamisha: Integral community activation via community-based tourism. In E. Mamukwa, R. Lessem, & A. Schieffer (Eds.), *Integral green Zimbabwe: An African phoenix rising* (pp. 67–86). Gower Publishing Limited.

Martins, C. (2011). The dangers of the single story: Child-soldiers in literary fiction and film. *Childhood*, *18*(4), 434–446. https://doi.org/10.1177/0907568211400102

Mason, C. L. (2016). Tinder and humanitarian hook-ups: The erotics of social media racism. *Feminist Media Studies*, *16*(5), 822–837. https://doi.org/10.1080/14680777.2015.1137339

McClintock, A. (1995). *Imperial leather: Race, gender and sexuality in the colonial contest*. Routledge, Inc.

McGehee, N. G., & Andereck, K. (2008). 'Pettin' the critters': Exploring the complex relationship between volunteers and the volunteered in McDowell County, West Virginia, USA and Tijuana, Mexico. In K. Lyons & S. Wearing (Eds.), *Journeys of discovery in volunteer tourism: International case study perspectives*. CABI.

Mkono, M. (2011). The Othering of food in touristic eatertainment: A netnography. *Tourist Studies*, *11*(3), 253–270. https://doi.org/10.1177/1468797611431502

Mowforth, M., & Munt, I. (2009). *Tourism and sustainability: Development globalisation and new tourism in the Third World* (3rd ed.). Routledge.

Moyo, S., & Tichaawa, T. M. (2017). Community involvement and participation in tourism and development: A Zimbabwe study. *African Journal of Hospitality, Tourism and Leisure*, *6*(1), 1–15.

Mudimbe, V. Y. (1988). *The invention of Africa: Gnosis, philosophy and the order of knowledge*. Indiana University Press.

Nelson, V. (2005). Representation and images of people, place and nature in Grenada's tourism. *Geografiska Annaler: Series B, Human Geography*, *87*(2), 131–143.

Pailey, R. N. (2019). De-centring the 'white gaze' of development. *Development and Change*, *51*(3), 729–745. https://doi.org/10.1111/dech.12550

Perkins, H. C., & Thorns, D. C. (2001). Gazing or performing? Reflections on Urry's Tourist Gaze in the context of contemporary experience in the antipodes. *International Sociology*, *16*(2), 185–204.

Phillips, K. R. (2002). Spaces of invention: Dissension, freedom, and thought in Foucault. *Philosophy and Rhetoric*, *35*(4), 328–344. https://doi.org/10.1353/par.2003.0010

Rice, L. (2003). Of heterotopias and ethnoscapes: The production of space in postcolonial North Africa. *Critical Matrix*, *14*, 36–75.

Rideout, L. (2011). Representations of the 'Third World' in NGO advertising: Practicalities, colonial discourse and western understandings of development. *Journal of African Media Studies*, *3*(2), 25–41. https://doi.org/10.1386/jams.3.1.25_1

Saarinen, J. (2011). Tourism development and local communities: The direct benefits of tourism to Ovahimba communities in the Kaokoland, Northwest Namibia. *Tourism Review International*, *15*, 149–157. https://doi.org/10.3727/154427211X13139345020534

Sadoma, Z. W., Chigwanda, B., & Manyati, T. (2015). Education in the crucible of a postcolonial revolution: Community reaction to the challenges of Zimbabwe's crisis. *Social Dynamics*, *41*(2), 235–252.

Salazar, N. B. (2009). Image or imagined? Cultural representations and the "tourismification" of peoples and places. *Cahiers d'études Africaines*, *49*(193–194), 49–71. https://doi.org/10.4000/etudesafricaines.18628

Schwarz, K. C., & Richey, L. A. (2019). Humanitarian humor, digilantism, and the dilemmas of representing volunteer tourism on social media. *New Media & Society*, *21*(9), 1928–1946. https://doi.org/10.1177/1461444819834509

Simoni, V. (2014). From tourist to person: The value of intimacy in touristic Cuba. *Journal of Tourism and Cultural Change*, *12*(3), 280–292. https://doi.org/10.1080/14766825.2014.934054

Simpson, K. (2004). 'Doing development': The gap year, volunteer-tourists and a popular practice of development. *Journal of International Development*, *16*(5), 681–692. https://doi.org/10.1002/jid.1120

Smithers, K. (2023). Heterotopia and the 'image of Africa': School tours and philanthropy in a Zimbabwean school. *Critical Studies in Education*, 1–17. https://doi.org/10.1080/17508487.2023.2286230

Staples, A. J. (2006). Safari adventure: Forgotten cinematic journeys in Africa. *Film History: An International Journal*, *18*(4), 392–611.

Sun, J., & Yanjun, X. (2020). The 'internal orientalism': New encounter in Tibet tourism. *Current Issues in Tourism*, *23*(12), 1480–1492. https://doi.org/10.1080/13683500.2019.1658724

Suzuki, Y. (2017). *The nature of whiteness: Race, animals, and nation in Zimbabwe.* University of Washington Press.

Taru, J., & Gukurume, S. (2013). Corporate responsibility in the tourism industry: Lessons from communities surrounding Great Zimbabwe monuments. *Journal of Tourism Challenges and Trends*, *6*(1), 59–74.

Taylor, J. T. (2015). "You can't spend your whole life on a surfboard": Elvis Presley, exotic whiteness and native performance in Blue Hawaii and Girls! Girls! Girls! *Quarterly Review of Film and Video*, *32*(1), 21–37. https://doi.org/10.1080/10509208.2012.757532

Thompson, J. (2022). Volunteer tourism fields: Spaces of altruism and unsustainability. *Current Issues in Tourism*, *25*(5), 779–791. https://doi.org/10.1080/13683500.2021.1887824

Toomey, N. (2017). Humanitarians of Tinder: Constructing whiteness and consuming the Other. *Critical Ethnic Studies*, *3*(2), 151–172.

Tucker, H. (2019). Colonialism and its tourism legacies. In D. J. Timothy (Ed.), *Handbook of globalisation and tourism* (pp. 90–99). https://doi.org/10.4337/9781786431295

Urry, J. (1990). *The Tourist Gaze: Leisure and travel in contemporary societies.* Sage Publications.

Urry, J., & Larsen, J. (2011). *The tourist gaze 3.0* (3rd ed.). SAGE Publications.

van Beek, W. A. (2003). African tourist encounters: Effects of tourism on two west African societies. *Africa: Journal of the International African Institute*, *73*(2), 251–289. https://doi.org/10.2307/3556890

van der Duim, R., Peters, K., & Wearing, S. (2005). Planning host and guest interactions: Moving beyond the empty meeting ground in African encounters. *Current Issues in Tourism*, *8*(4), 286–305. https://doi.org/10.1080/13683500508668219

Yapa, L. (1996). What causes poverty?: A postmodern view. *Annals of the Association of American Geographers*, *86*(4), 707–728. https://doi.org/10.1111/j.1467-8306.1996.tb01773.x

ZIMSTAT, & World Bank. (2020). *Zimbabwe poverty update 2017–2019.* https://www.zimstat.co.zw/wp-content/uploads/publications/Income/Finance/Zimbabwe_Poverty_Updat_2017_19_Final.pdf

# 6 The starring role

## Poor-but-happy children

In this chapter I detail the discourses associated with the children at the school, or the actors in the metaphor. On the movie set, the children are expected to conform to the image of the 'poor-but-happy' child, in which they must always be smiling and ready for a picture. I begin by outlining the role of the poor-but-happy child. The interviews and observations revealed that the poor-but-happy discourse is vital for sustained funding of the school. For this tourism to be successful, the role of the poor-but-happy child is a critical stereotype for the tourists to witness, as it allows them to feel comfort in their visit to the school. If the children were unhappy, the tourists may have to problematise their own implication in the level of poverty they witness. By providing an image of poverty in which the participants are happy, the tourists can ignore the lack of material assets by arguing that the people are happy and consequently material assets must not be of importance. Their role in this metaphor may empower the tourist to play the role of the white saviour (see Chapter 7).

I then outline the 'love' the tourists are reported to have for the children, its discursive basis, and the reaction of the teachers to the way their relationship to the children is constituted. The teachers' opinions are important here, as they are the directors of the film in which the poor-but-happy children are the main characters. In the section, I detail the naming of the tourists in the school in which I examine the ways in which the tourists are exclusively called 'visitors' when they are in the school, and they are called 'friends' when they are holding hands with children – an interesting component of the tourism that I argue is a deliberate strategy.

### Playing a character: poor-but-happy children

The trope of the poor-but-happy child alleviates some of the discomforts a Minority World tourist may feel witnessing poverty, as it allows them to excuse the living conditions of the people that they are viewing (Crossley, 2012; Frazer & Waitt, 2016; Simpson, 2004). The tourist can excuse the conditions by identifying that the participants are happy, so their living conditions must, therefore, not be so bad. Frazer and Waitt (2016) describe the poor-but-happy child

DOI: 10.4324/9781003407980-6

representation as a 'discursive romanticisation of poverty' (p. 184) which governs the actions of those who partake in it and enables them to ignore the uneven relationships from which they can gaze on the poverty.

During school tours, I often heard exclamations from tourists about how happy the children were, for example, in an extract from the field notes:

> I walk into the Grade 5 classroom and overhear, 'I love how happy they are, at a normal school they are not this happy'.

The children also noticed the praise during the tours. When asked what conversations they would have with tourists, they said the following:

> They say, 'This is nice classroom' or they say, 'This is nice children'.
> (Brenden, age 10)

> About the school? They will say, 'The school is nice'.
> (Gracious, age 11)

The construction of 'poor but happy' creates conditions for poverty to be exotified, in which the differences between the host community and the tourist are emphasised (Simpson, 2004). Further, the image of the smiling child reinforces the idea that the saviour/donor/tourist are performing a 'good' act, as the children appear to want them there. In this construction, poverty may be different to the poverty experienced in the tourist's home country. The assumption in the quote above, that Matopo School is not a 'normal' school, is an exotification of the school and the children in it. Aitchson (2001) argues that 'the construction of the Other is dependent on a simultaneous construction of the "same"' (p. 137). Tourists use their construction of 'normal', their vision of an American school, to explore the differences they are witnessing on the school tour. Othering is, and always has been, a dominant discursive framing of tourism, as tourists seek cultures which are different from their own (Lozanski, 2010b). In the comparison of Matopo School and a 'normal' school, there is a fantasised difference, which juxtaposes the Other against the Minority World norm. In this case, happy children in the school are considered at odds with the experience of American children. Implicit in this framing is the notion that the children are happy as they appreciate the 'little things' rather than being wedded to capitalistic notions of happiness (Lozanski, 2010b).

The discourse of poor-but-happy was pervasive in both the field notes and in the interviews. For example, Alison, a school founder, describes the development of the children at the school over time:

> The children back then were [**pause**] were incredibly shy. As a white person meeting them it was like you were a ghost! I mean they were terrified of me when I first appeared. And they were just awkward, really surly, you would not get a smile out of them at all. And you just would not

recognise the kids now, as being the same kids then. And yes, most of them were younger but to see over just a few years these kids have absolutely blossomed. In confidence, in smiles and laughter.

(Alison)

The discourse of the children becoming happy and full of smiles is part of the continuing subjugation of children in the Majority World (Bell, 2013). Further, the construction of children as happy flattens the image of the child, to that of a person who is only able, or allowed, to have one emotion: happiness. It is a taken-for-granted assumption that functions to govern the relationships between the children and tourists. Schools are a medium through which bodies are always-already controlled, regulated, and disciplined (Foucault, 1975/1991). In *Discipline and Punish* Foucault (1975/1991) developed the idea of a technology of power capable of manipulating and governing bodies, 'a policy of coercions that act upon the body, a calculated manipulation of its elements, its gestures its behaviour' (p. 138). In Alison's construction of the children, the children only 'blossom' when they begin to fit *her* image of what a child, or childhood, should be: happy and smiling. The image of a 'happy' child as the norm governs the image of childhood, which is distributed through the Minority World. The discursive construction of childhood in the Minority World imaginary creates conditions of possibility of what a child can be, how they should act, or how they should look, which in turn privileges Minority World conceptions of childhood. It is a vision of childhood which examines children in the Majority World in terms of what they 'lack' when compared to the dominant discourses of childhood in the Minority World (Aitken, 2001).

These representations mark Majority World childhoods as Other to modern Minority World childhoods, which in turn enables a dominant discourse of childhood as one which should be filled with play, innocence, and happiness (Carpenter, 2015). Kesby et al. (2006) argue that in Zimbabwe there are hybrid childhoods which are reflections of 'the nation's complex historical geography' (p. 187) and thus any representation of childhoods in Zimbabwe should reflect this hybridity of childhoods, rather than encourage Minority World notions of childhood. In this case, Alison's understanding of childhood does not recognise the hybridity of childhoods, and instead encourages a Minority World representation of an African childhood. A binary understanding of childhood reflects Foucault's dividing practices, which enable the judgement of correct and incorrect ways of being. As Foucault (1975/1991) described, the dividing practices:

makes possible a whole investment of duration by power: the possibility of detailed control and a regular intervention in each moment in time; the possibility of characterising, and therefore of using individuals according to the level in the series that they are moving through.

(p. 160)

There is a dominant image that governs Alison's understanding of the school, and how the children should behave within it. Further, this discourse makes possible forms of regulation and governance, as children in the school must strive to meet the dominant image of the 'happy child'. A discussion of the aestheticization of poverty and image of the Other in the Minority World is not necessarily new (Mowforth & Munt, 2009). The image of the poor-but-happy child is captured by the tourists through the lens of their cameras and mobile phones, allowing this image – and the associated discourses – to continuously circulate globally through social networking websites. The photography of the children itself calls into question several ethical concerns about this type of tourism, as every visit was accompanied by multiple photographs being taken. Hernandez (2018) details the troubling function of photography:

> Photographs can often, then, function as distorted forms of representation and appropriation of subaltern bodies. As such colored bodies are depicted, for example, as smiling faces that portray the acquiescing oppressed as happy with their placement within the colonial matrix of power.
>
> (p. 192)

Crossley (2012) experienced the same phenomenon in their research regarding a volunteer tourism trip, in which the majority of their group began photographing young children at a school before they had been introduced and did not ask for consent. The use of photography allows a perpetuation of both the discursive construction of the poor-but-happy child and a heterotopic image of Africa. As Staples (2006) identifies, photography and film has been critical to the construction of the 'imaginary geographies and global touristic subjects who have maintained certain ideological, social, economic and political relationships with peoples and places visited' (p. 407). The colonial relic of photography to capture the exotic Other ensures that the use of photography for the tourists is an uninterrogated act which furthers an image of Africa in the Minority World imaginary.

The uninterrogated act of photography and the subsequent distribution through social media is explored by Toomey (2017) in their study of the website *Humanitarians of Tinder*. *Humanitarians of Tinder* is a website through which people anonymously submit screenshots of Tinder profiles that contain images of the Other – problematic in itself, as it furthers the dissemination of images that may not have permission to be taken. Toomey argues that there can be specific motivations behind the photographing of people in Majority World countries,

> Posing with, and then posting one's interactions with, the Other is a form of consumption, whereby the body of the Other decorates the photo's subject with the goal of making the subject seem more interesting, thus using this interaction with the Other as a type of social capital.
>
> (Toomey, 2017, p. 158)

The continued use of people in the Majority World to provide positive social recognition for people in the Minority World is furthered by dissemination of pictures which include the poor-but-happy child. Returning to the metaphor, the role the children play in this movie is one that is eclipsed only by one other, that of the white saviour – who plays the protagonist. In their role the child must remain happy, as to enable the white saviour to ignore the reality of the poverty they may witness. This role positions the child as the happy receiver of a gift and allows the white saviour to alleviate their own guilt about their relatively immense privilege. Any gifts given to the children reassures the white saviour that they are helping to 'make a difference' to the lives of the children. In turn, this gift exchange is photographed and disseminated via social media networks of the white saviour.

The image of the poor-but-happy child perpetuates a silencing of the structural reasons for poverty and supports the well-worn argument of the world as a meritocracy, which suggests anything is possible if you work hard enough. An ignorance of poverty is discursively linked to the dominant discourse which constitutes children in the Majority World as innocent, and therefore not to blame for their own poverty (Sheridan et al., 2019; Simpson, 2004). The poor-but-happy child image ensures that the tourists can temporarily forget the world economic systems which create and maintain extreme poverty. A forgetting of the broader structures which cause intergenerational poverty was present in some of the field notes, with an incident of inspiration detailed,

> A woman from America stands up and she explains that she had come from a rural town in America where no one usually went to university. She explains how she became a doctor and then concludes as she says to them 'If you try your hardest and give 10%, someone will come along and give you the other 90%.' What sort of message is that to send? Just try only a little bit and you will succeed? 10% would not even be a pass mark in school, not to mention there are plenty of graduates in Zimbabwe who do not have jobs, despite working hard.

The inspirational message espouses individual accountability, which is common in discourses in the Minority World. The suggestion that children at Matopo School can work 10% and someone else will provide the other 90% ignores the structural reasons, such as the economy, which are the cause of poverty in Zimbabwe. The reality for most in Zimbabwe is due to economic turbulence; even if children work hard, there may be no employment prospects that match the qualifications they obtain – some report 95% of the working population in Zimbabwe are earning a living from the informal sector (Mhazo & Thebe, 2020). Despite best efforts to inspire, the children at Matopo School are setup to fail in a system that continues to privilege urban, usually white, upper-class citizens. Further, despite hard work at school, many children in Matabeleland North still fail their O-level exams; with the region having a high failure rate (ZIMSTAT, 2017).

The inspirational message provided to the children represents the white saviour fulfilling their script of inspiring the poor-but-happy child, by providing them with hope rather than material assets. There is a discourse of innocence associated with Majority World childhoods, which Carpenter (2015) argues 'helps foreigners to reconcile the dissonance they experience when their internalized ideology of individual accountability is confronted with Majority World poverty for the first time' (p. 20). The poor-but-happy child reflects the neoliberal discourses in the Minority World of individual accountability as a reason for success, rather than historical or structural factors. These discourses protect the white saviour from the realisation that their individual actions cannot make a real difference in the lives of those in the school, as there are fundamental economic world structures at play which perpetuate poverty in the Majority World.

Despite the assertions by the tourists that the children were unusually happy; like all children, the children in this study expressed multiple emotions throughout the course of the school day. In Zimbabwe, schooling is often a serious endeavour with Greg, a tour guide, explaining,

> The education system here is incredibly conservative. And, you know a lot of the discipline and the way they do things, you might as well be back in the 1960s or 1970s. In the American system you are taught to question everything! Here, it is, no. Deh deh deh [**indicating following a set pattern**]. And that is how it is.

Thompson, the school founder, also explained how schooling in Zimbabwe was serious and identified how he wanted his classroom teachers to have more fun with their students,

> The teachers are used to the seriousness of teaching and that you have got to be the authority. I am like, 'you have to have fun with these kids' so that they respect you. Because otherwise they will have this negative thing about school. I have got a completely different way of looking at it. The teachers think there is no time for fun. That is what I hate, in class where you know, there is not a joke the whole time. You know, it is no fun, they just think that school is regimented and that is not good.

Both Thompson and Greg describe how school has a serious academic focus in Zimbabwe. For Thompson, the happiness of the children at his school is highly unusual for a school in Zimbabwe,

> And the kids, have just, I do not know who tells them but I have never been greeted so many times in one day – 'hello', 'good morning Mr. Thompson' and [pause] I do not know if the teachers do say, 'you have to be nice to Thompson!' or they just know that this is the person who is helping us, I do not know. It is just amazing. I do not know it is just [**pause**] I am finding it really, really lovely that the kids are so friendly and learning well. They were not able to say a word of English at the beginning.

In fact, the 'happiness' of the children is such an abnormal behaviour that he believes somebody must have told the children to be nice to him.

The children were aware of the need to present a particular image in order for the tourism to continue. It is possible the children engaged in strategic behaviour as they knew the benefit tourism had for them. While children in their interviews identified that they 'liked' the tourists visiting, often the reasoning was that the tourists brought gifts for the children:

> I would prefer to come to school which has visitors because they will be helping us. When there are no books, they will be paying, they will, be giving us some money, and we buy some books.
>
> (Tawanda, age 10)

> With visitors is coming here to school. With the visitors you take your photo. Visitors give all books, pencils, pen.
>
> (Lovely, age 10)

These extracts are contradictory to the image of innocent children who are overcome with joy, bursting into singing and dancing and suggests the children know the reason and payoff of tourist visits (for futher exploration see Smithers, 2022). These extracts suggest that the children show joy and happiness about the tourists due to the potential payoff: gifts. The children did enjoy the visits, but it may not necessarily be due to an inherent love of dancing and singing for the tourists.

There is a disjunction between what the tourists saw as 'happiness' and the children's experience in the every day. Field notes I wrote during this time explore this disjunction,

> I wanted to say they are not always happy! I always have children crying to me about something!

As most teachers of children who are in primary school could attest, children have arguments, they cry, and they have emotions. In the interview with Greg, a tour guide, he also identified that the way that the children reacted to the presence of tourists was different to those at other schools,

> And you will see they [other school in the area] just do not get visitors. So, the kids might be a lot more shy, they will be a lot more... they will not come up and say 'hi, how are you?' you know. At Matopo School they do, and they are encouraged to do that, and you will see the kids actually fighting to get hold of *their* tourists.

It is understandable that both Thompson and Greg felt that the children being happy was unusual, as most children will be a combination or spectrum of happy to upset within the day. The ranges of emotions children display, coupled with the academic and traditional nature of schooling in Zimbabwe, means that for *all* the children to be happy and smiling during a two-hour period – like that of the tour – is unusual. Yet, this image persists in the narratives about images of Africa. The discourse of the happy child shapes the expectations of the tourists, and in turn the subjectivities of the children.

The teachers are aware of the role that the children must play in this metaphor. As directors of the movie, the teachers work to produce a heterotopia built from the image of Africa for the tourists in the school, as discussed in the previous chapter. The teachers play the role of movie director, as they oversee the entertainment and govern them to ensure they conform to the image of a happy and smiling child. One example of this is the presence of a teacher on all tours, despite this providing many interruptions to the school day. When asked about this, many teachers said it was to help the children with their English if required or to stop children from saying incorrect things to the tourists. During the field work there was an incident which highlighted one of the components of the image the school was trying to portray:

> Tabitha gave a long speech at the teachers' meeting this morning about teachers performing their duties correctly. She said on Thursday, when the Education Tour group visited, two children in Grade 1 had been having an argument. This culminated in a physical fight which the prefects, who were assigned to the class, could not end. Eventually one of the guests from the tour group had to tell them to stop. It was considered unacceptable for this to happen, as the children must always present smartly to guests.

On this day, there was very few teaching staff present at the school, for various reasons. When an argument occurred between two of the children, there was no adult supervision to prevent the tensions from escalating into a physical fight. Tabitha's focus at the teacher meeting was on the unacceptable nature of this incident occurring in front of the tourists. Rather than annoyance about the physical fight occurring, the frustration is about the incident occurring *in front* of the tourists and that the tourists had to stop the argument. When the tourists witnessed the argument, the argument has the potential to disrupt the dominant discourse, as it might suggest that the children are in fact not happy all the time. Further, to make use of a tourism term, the 'backstage' of the school tour was under threat of becoming exposed. There was a risk that a tourist might see something other than a happy child. In the incident described above, Tabitha is trying to ensure that there will be no future altercations which might suggest that the children are anything but happy. If the children were to appear anything but happy, their subject positioning as innocent and happy may be revoked – and as a consequence, the perceived immediate need for funding and Minority World intervention may also be revoked (Bandyopadhyay & Patil, 2017).

Whilst tourism scholars often use the term 'backstage' to represent the actions that occur behind the scenes to maintain tourism destinations, the work of the teachers and student is also subject to a surveilling tourist gaze (Minca, 2010). A surveilling gaze turns surveillance into a normalised practice which allows for self-governing of behaviour,

> he who is subject to a field of visibility, and who knows it, assumes responsibility for the constraints of power; he makes them play spontaneously upon himself, he inscribes in himself the power relation in which he simultaneously plays both roles: he become the principle of his own subjection.
>
> (Foucault, 1975/1991, pp. 202–203)

Here, I have explored the construction of the poor-but-happy child and interrogated the taken-for-granted assumption that the children are happy all the time. I argue that the children are aware of the image they need to play and they conform to this image. The children and teachers then engage in self-governing behaviour.

In the movie metaphor, the children are actors in a movie fulfilling a script. It could be argued that in many ways the dance and tour is a *real* performance, outside of the metaphor, and it is produced by the power relations which work to govern and circulate the image of the poor-but-happy child. I now explore a discourse regarding the tourists, with this dominant discourse positioning tourists as having considerable affective associations with the school tour and in the relationships they form with the children.

### 'They are in love with the children'

The poor-but-happy child trope is at work in Matopo School, and there are taken-for-granted assumptions which work alongside it. There is a reverse side of the same coin: the 'love' that the tourists are reported to have for the children, which I now explore. Whilst tourists envisage the children as happy, the school constitutes the tourists as having a love for the children.

In relation to cultural tourism, Lozanski (2010a) offers a powerful assessment of the relationship between intimacy and the exotification of the Other:

> The possibility of being 'loved' by these Others is contextualized by a broad infrastructure that imposes difference – and distance – through discrepancies of racialization, class, nationality, and mobility. Travel in particular is underpinned by a sense of colonial entitlement in which Other cultures are objects for travellers not only to observe, but to know and understand. Distance and desire operate in tandem as a means through which to know, and be known by, the Other without compromising the structurally embedded colonial status that makes this knowledge and desire the purview of travellers.
>
> (p. 747)

As Lozanski further explores, many tourists seek 'intimate' encounters which they believe are an 'authentic' representation of a place.

Seeing and interacting with children is a key element of the tour. Greg, a tour guide, describes how it makes the tourists feel:

> And they are always very touched when they go out there and they see the kids and all that.
>
> (Greg)

For Greg, one of the most important elements of the school tour is the inter-action between the children and the tourists. Tourist expressions of joy and willingness to see the children was also mentioned by many of the teachers at the school. Zibusiso encapsulates the image of the tourists by the school,

> Because you see, some of them they have got so much love, they are in love with the children. When they come here, they do not want to leave, yeah, they take too much time to interacting with the children.
>
> (Zibusiso)

Zibusiso describes the tension felt between allowing the tourists to interact with the children and the interruption to the school day (as explored in the previous chapter). The protective discourses which constitute children in the Majority World as needy are embedded in the image of the poor-but-happy child. Judge (2015) suggests 'the fact that the emotional narratives often involve interactions with children is significant' (p. 165). Feelings of love could be con-sidered a hark back to colonial power relations through which parent-child discourses were used as imagery to support colonial paternalism (Judge, 2017; McClintock, 1995).

It is well-established in postcolonial studies that equating colonised people with children was a tactical position to justify the cultural dominance of Europe (Fanon, 1986). Mills and LeFrançois (2018) argue,

> Populations in the global South are not only being talked about as chil-dren, they are being acted upon as if they were children, with global North countries working *in loco* parentis for children of the global south... especially those in receipt of aid.
>
> (p. 517)

The parent-child relation draws the tourist into a number of complex and contradictory positions, as they navigate between being an interloper, a valued guest, an enamoured parent-like-figure and the white saviour. The relation between tourist and child was documented in the children's draw-ings. Most included a tourist holding a child's hand, as demonstrated by the drawings from Precious (age 10) and Shama (age 9) (Figures 6.1 and 6.2).

*Figure 6.1* Drawing by Shama depicting tourists standing under a tree, asking the child questions.

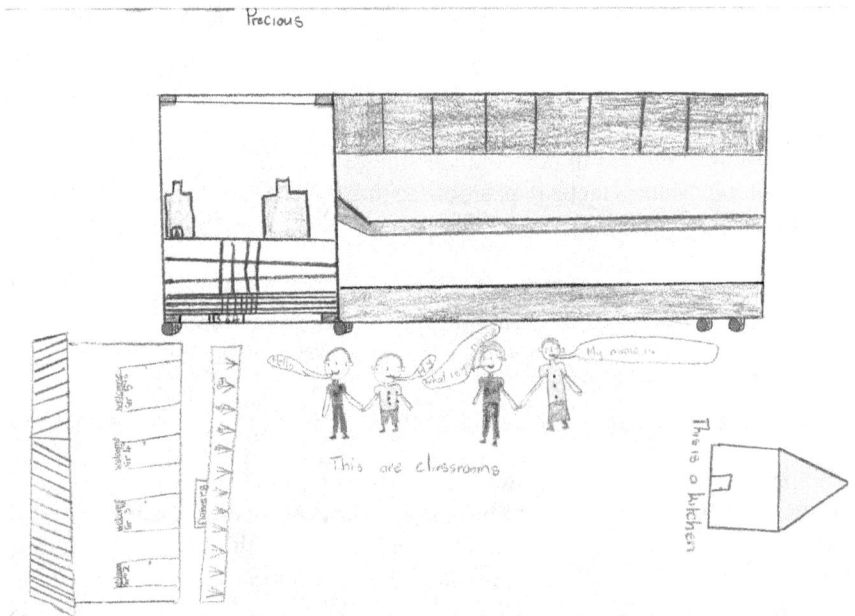

*Figure 6.2* Drawing by Precious depicting a school building, a school bus, and tourists asking children questions.

In the drawing by Shama, the tourists and children are depicted as standing underneath a large tree, with the school building in the background. On the left a tourist holds a child's hand, and asks 'what is your name?'. In the drawing by Precious, the school building is depicted on the left hand side, the kitchen on the right, and the tourist bus in the centre. There are four figures, with two tourists holding a child's hand each. One tourist says 'hello' and the other asks 'what is your name'. As I have discussed elsewhere (Smithers, 2022), the children had a very one-sided interaction with the tourists, as they were asked repetitive questions over and over with each tour group. The image of loving and caring tourists was not present in the interviews with children or in their drawings.

As Zibusiso details above, one of the dominant constructions of the tourists by the teachers is of a love, care, and enjoying the children's presence. When asked how the tourists would react to not being able to interact with the children, Zibusiso said,

> Because I think it will be personally unfair to say, when our guests come here, we say just walk around with one guy guide, yet they come to a school and they want to see the children. We will be depriving them – even the assistance they will not give as much as if they interacted with the children.
>
> (Zibusiso)

Zibusiso recognises that if the children were not included in the school tour, the tourists may not donate as much money/goods. In the school teachers' construction of the tourists, kindness is directly related to involvement with the children, and this involvement is a determinant of the income stream for the school. In this construction, it is impossible to imagine a form of this tourism which does not include an interaction with the children. As such, this discourse constrains the manner the tourism *could* operate in the school. The construction, of needing children to maximise funding, governs the implementation of tourism in the school. Further, the school tour creates a normalising gaze through which the children must conform to the 'fun' and willing child subject, who will respond positively to a tourist who wants to give them care and love.

Greg, a tour guide for Education Tours, also felt that the tour would be perceived differently if there were no children included in the tour. When asked about removing the high level of interaction from the tour he said,

> No, I think that you have got to have the interaction. I think it would be a lot more flatter as a tour, and I do not think they would be as generous if they did not have the kids there. Because it is an experience, it is a real touchy feely, 'aww how are you doing, I am so and so and this is where I come from and where do you come from and… how old are you,' and that sort of thing and it is, you know, they enjoy that and they want to see that.
>
> (Greg)

In fact, Greg describes it as potentially *flatter as a tour* and that the tourists will not be as generous. He describes the experience as *touchy feely* and that the tourists *enjoy that*. The tourism has an affective component which relies on the development of tourist perceptions of love and care. Greg works for Education Tours, who allow the school to have a large level of control over how the tour is run in the school. Despite Matopo School having a level of freedom from Education Tours to run the school tour in a manner they choose, Greg identifies that children are a vital component for the tourism. The narrative that the children are *vital* reinscribes imagery and discourses associated with innocence and dependency. Discourses of innocence and dependency are insidious in volunteer tourism encounters (Judge, 2015; Mostafanezhad, 2013a), and as outlined here, are insidious in developmentourism and school tours. The narratives of dependency and innocence are significant to explore as they echo colonial power relations.

Mary, a tourism employee who was involved in distributing funds from Adventure Company, also believes that children are vital to funding efforts. When asked whether children were needed for the tourism, she said,

> Yes, I think, you can have a little face there and people are conditioned to think, 'oh protectiveness dah dah dah'. But when you look at an adult, they actually think, 'come on make a plan'. But they do not realise, and that is, where even yesterday, I was trying to show the relationship between teachers and pupils. You do not have a school if you do not have those teachers, so you have got to look at that.
>
> (Mary)

In this extract she identifies the difference between the dominant subject positioning of children in the Majority World and the dominant subject positioning of adults in the Majority World. For children, whose innocence in this subject position is assumed, help is *always* needed and deserved (Liebel, 2019). For adults, the solution is to *make a plan,* overcome their problems and 'fix' themselves out of poverty (Wale & Foster, 2007). Mary recognises the subject positioning of children as innocent, which is the normative gaze through which tourists encounter children in the Majority World. In the second half of the extract, Mary details the problems she encounters when trying to gain funding for projects related to the teachers. Part of the complexity of this tourism is that in the case of helping a school, often just helping the children is not enough.

Thompson, a school founder and tour guide, invokes the discourse of 'naturalness' to explain why children are preferred:

> It is just the way people perceive things; it is just natural. Yeah. If you took a picture of the little baby elephant and posted it, it is looking cute and all. You will have much better comments, then if you took a big elephant just standing there in the shade.
>
> (Thompson)

Mary and Thompson perceive the role of children in this type of funding as inevitable as people are conditioned to have feelings of protectiveness towards children. The 'conditioning' or feelings of it being 'natural' exemplifies how certain discourses become embedded in power relations. Foucault (1976/2008) identified of power relations, 'it is often the case that no one is there to have invented them, and few who can be said to have formulated them: an implicit characteristic of the great anonymous, almost unspoken strategies which coordinate loquacious tactics' (p. 95). It is in this 'almost unspoken' strategies that the idea of being 'conditioned' or 'natural' are at work – in the taken-for-granted assumptions which govern. Returning to the metaphor of the movie script, in their position as producers of the tourism, Mary and Thompson are in a position of influence and could begin to dismantle these taken-for-granted assumptions; however, in this case, the discourse of protecting children is incredibly powerful and embedded in the power relations which govern both the implementation of the tourism and those who work in the tourism.

Nolly identified that Adventure Company, the company Mary works with, had requested for the children to be *more* involved in the tours:

> But with Adventure Company they had to request that they want to be guided by the children, because they want more interaction with the children, and that is what is happening now. I think they are happy about that. Because, in particular, when they are coming here at school, their main purpose is to interact with the children. Yes, so I think when they are with the children, they are more happy.
>
> (Nolly)

The request to be guided by children is reflective of the tensions of implementing developmentourism and school tours – the tension of balancing the educational needs of the children versus the potential for further funding. The identification that the children are the main purpose for the visit, raises questions about if the visit needs to occur at a school at all. For example, are the tourists visiting to see how a Zimbabwean school works? Or, are they visiting to take a photograph with the poor-but-happy child they have seen in advertisements? Previous research would suggest the conditions of possibility for thinking about the visit are constrained by discourses of care, which position the visit within an affective need to 'give back' to innocent children who need intervention from adults in the Minority World rather than to visit a school to see how a school runs (Bandyopadhyay, 2019; Bandyopadhyay & Patil, 2017; Judge, 2015; Mostafanezhad, 2013b).

The focus on children was a somewhat contentious subject among the teachers. At the conclusion of every interview, I asked, 'is there anything you wish to add that we have not discussed?' and during this time most of the teachers raised the issue of the donations focussing solely on the children, rather than the school:

And most of the things when they are donating things, it must be children, children, children. No mention of the teacher... But I am busy canvassing for other children's fees to be paid for. They are benefiting a lot, by my being here, through seeing my services. But no one will notice to say, 'ahh can we take up this, can we also assist the teachers in doing this or if they have got children, can we also pay for their fees?' Or someone is saying, 'okay let me pay for two teachers?' And then maybe that will fix and improve our lives, because as it is, our life is just stagnant we are not moving upwards we are not moving backwards, we are just in same position where we are.

(Zibusisio)

Zibusiso's comments reflect the unequal power relations involved in the implementation of developmentourism and school tours. In one sense, Zibusiso's comment reflects the ways in which power is productive. The power relations within developmentourism produce the ideal teacher at Matopo School: the teacher who works hard whilst reducing disruptions caused by tourism (as identified in the previous chapter), but also the teacher who works to produce the correct image of the school, as to maximise funding for the children. Power is, 'a relation in which one guides the behaviour of others. And there's no reason why this manner of guiding the behaviour of others should not ultimately have results which are positive, valuable, interesting and so on' (Foucault, 1980/1988, p. 12). In this case, the power relations produce the 'good' teachers as the ones who do not question the lack of attention paid to their needs, something which is not unique to Matopo School or the Zimbabwean schooling system. In Matopo School, the production of the 'good' teacher is partially framed through colonial discourses of the innocent poor-but-happy child, but also through discourses that frame adults as needing to 'make a plan' for their own survival; rather than relying on support from others (Wale & Foster, 2007). In a Foucauldian sense, Zibusiso's subjectivity is constituted in relation to the discourses which frame a 'good' teacher at Matopo School – which form a complex picture of what the 'good' teaching professional should be.

Zibusiso was not the only teacher to speak regarding the lack of attention paid to teachers:

Wishing, that maybe even us as teachers in future, to come, maybe since we are receiving too much visitors, at our school. We are looking forward to one well-wisher who will come and say, 'No, I am here to also look at the teachers' requirements'. Like now we are studying, you will be studying, maybe the salary that you are getting is not enough for you, you are working for a family, you need to pay your own school fees, need to pay the kid's school fees, sometimes you want to start, you want to research you do not have a laptop, you are using a phone, now you think, 'with this little money, I cannot afford to buy a laptop'. Yes. So, we are just looking and wishing for the best, that a tourist continues coming at our school, one of the days we are going to find a well-wisher who is going to look at us teachers.

(Bridget)

The positioning of the tourists as solely focussed on the poor-but-happy children also creates a subjectivity which positions the tourists in a particular way. The tourists are constituted as focussed on children, rather than the holistic needs of the school. Thompson, a school founder, highlights this:

Well, there was a substantial amount donated towards the clinic, if you ask the same for the school you would probably get nothing. That is the difference. There is our other issue, make a campaign for teachers' salaries you will get zero. You will automatically get zero. Make a campaign for kids' fees, you will get some. So, what we do, play around amount of kids' fees, so that you can reserve some for paying the teachers. It is just... you have to work with the figures.

(Thompson)

In the 'donor-obligation nexus' there is a relationship created which is framed through discourses which, 'bind the grateful receiver into a nexus of obligations and duties towards the generous donor' (Chouliaraki, 2010, p. 113). The nexus of obligations might mean the school founders feel they cannot tell people how to donate, as their position within the obligations is to be a 'grateful' recipient. Thompson identifies the ways in which the school resists the determination that children should be the only beneficiary of the tourism. The school recognises the discourse which constitutes children as needy, innocent, and deserving, and they have artificially inflated the school fees required for every child. This means any child who is sponsored, or pays school fees, contributes towards helping to pay teacher salaries.

The poor-but-happy discourse is constructed through feelings of love and care. The feelings of love and care are framed through discourses that constitute children as needy and innocent. On the other hand, this discourse frames teachers as not in need or innocent (this idea was also explored in the previous chapter). As such, the school founders and teachers utilise the dominant discourse to maximise funding for the school. It was made clear, in interviews and observations, that the affective notions of 'love' and 'care' for children are one of the main reasons for donations to the school. In the next section, I detail the naming of the tourists in the school and the ways in which the teachers, as directors of the movie script, utilise this naming to their benefit.

### Tourist-visitor-friend

There was an intentional naming of the tourists by the school. Rather than simply referring to them as 'tourists', the school had particular discursive functions regarding the naming and greeting of the tourists. The naming is explored here as the teachers describe the way in which the school community must conform to discourses of hospitality and care, due to this naming.

During the fieldwork it became apparent that the teachers usually referred to the tourists as 'visitors', rather than tourists. As part of the reflexive nature of qualitative research, I began to ask the teachers about the usage of this term (see e.g. Brinkman, 2018). When asked, most of the teachers were perplexed by the question regarding the naming of the tourists and took a moment to consider their answer. For them, it was an uninterrogated way of speaking. It is in the taken-for-granted and uninterrogated ways of doing and knowing that discourse can serve a 'tactical function which is neither uniform nor stable' (Foucault, 1976/2008, p. 100) and 'can be both an instrument and an effect of power' (p. 101). In fact, this uninterrogated way of speaking was incredibly powerful, and I took on this naming during the course of the fieldwork, most of my field notes refer to tourists as visitors. To this day, I still refer to the tourists at Matopo School as visitors in conversations about my research. When I returned home to Australia and begun to consider the interviews with teachers and school founders, it became apparent there was a second term, 'friends', which was also used to describe the tourists.

For some teachers the naming was part of the established school routine. Ingrid identifies how she followed the lead of other teachers when she joined the school,

> Visitors. [laughs] I do not know. I think that is the arrangement [both laugh]. I think that is the arrangement. Because I have been there for a short time so maybe they have their own arrangement, that I do not know.
> (Ingrid)

For Ingrid, the term is just another part of the ritual associated with the tourists visiting. In discussions of tourism there are arguments about 'staged authenticity', that is, there have been considerations as to how 'authentic' a tourism can be if there is an element of routine or script (Cohen & Cohen, 2017). The reflection by Ingrid suggests that calling the tourists a 'visitor' is an unspoken ritual within the school, rather than a deliberate attempt to create a feeling of care or relationship. In such, some tourism scholars may call this an 'authentic' encounter, through which the tourists and school community forge real relationships of care; however, it could be argued that the positioning of the tourists as a 'visitor', although unspoken, is part of the complex power relations which may help to constitute the tourists into becoming a 'donor', and in turn, completing the obligations of how a donor, or visitor, might look.

After reflection, some of the teachers concluded that if somebody was visiting with a 'positive motive', they could be called a visitor,

> What I understand is that when somebody comes with a particular positive motive, you cannot call them a tourist. Of course, they are coming to tour, but there is something behind the name tourists, but when we call

them visitors, it is like when the visitors are coming even to your home, they are a blessing. All what we have is a positive mind about them. I think that is why we are calling them visitors.

(Nolly)

Nolly identifies the main image of the tourists: they are visiting with a positive motive. She argues that visitors are similar to somebody who is visiting your home and refers to them as a blessing. She identifies *all we have is a positive mind about them*, arguing that the tourists are positive and calling them a tourist would alter the perception of them. It could be argued the naming has a specific purpose – allowing the children and teachers at the school to remember that the tourists have a positive purpose and should be treated with respect and care. The feelings of care for the visitors may be reflective of the 'love' the tourists have for the children, as discussed in the previous section. In the articulation, *all we have is a positive mind* there is an acknowledgement of the practices of gift-giving, in which through the act of providing gifts a moral agreement is made (Hernandez, 2018). As such, the school and the children govern their own behaviour in line with the moral code of accepting a gift. The moral code of accepting the gift allows the subjectivation of the children and teachers through which they become accepting and docile receivers of the gift.

Tabitha identified that calling the tourists a visitor has a specific way of considering the needs of a person and providing them with a very positive experience. She said that the term visitor reminds the children and teachers to treat the tourists in a different manner:

This is our visitor, we treat our visitor nicely, we show him around, make sure we give everything that that visitor needs. So, when he goes back [**pause**] he can remember us. We are saying, we are creating a second home for this person. To make them feel, we want to make them feel at home, so they are free to join in our activities and be with us, move around with us.

(Tabitha)

Tabitha identifies that the overall aim of treating the tourists positively is to create a connection, *a second home* and the aim is for the tourist to *remember us*. Here Tabitha hints at the main aim of the tourism in the school: to create a conduit of funding for future endeavours. For Tabitha, the naming has a specific purpose of creating a welcoming and positive environment for the tourists. The naming reminds the teachers and children that they need to treat the tourists with care to allow for future funding. Through a discourse of care, the naming furthers the subjectivation of the teachers and children. It governs their behaviour to ensure that they *create a second home* for the tourists and then make them feel at home in the school. In turn, the discourse of care enables and constrains certain ways of being, speaking, and acting. If the children and teachers are to act as if the school is a tourist's second home, the manner in which they engage with a 'visitor' will be different to the way of engaging with someone who is labelled a 'tourist'.

Developing a feeling of 'being at home' is part of the expectations associated with a discourse of 'authenticity' through which tourists are always seeking 'authentic' experiences (Burgold & Rolfes, 2013). By creating a 'home' for the visitor, the school aims for the tourist to feel that their visit is developing an authentic connection, something Cravatte and Chabloz (2008) label an 'enchantment' whereby, 'the relations between people are characterised by a state of euphoria that eliminates any feeling of falseness' (p. 237). The 'authentic' connection is part of the deliberate image the school puts forward as an attempt to continue tourism, the discursive construction of the 'visitor' governs the children and teachers by enabling and constraining the ways they can act and speak.

Zibusiso identified that they were called visitors, rather than tourists, as they know what to expect when arriving at the school and brought gifts for the children,

> I think they are visitors because when they leave the place they will be staying, specifically they know that they are going to a certain place unlike a tourist who be just going exploring the places but for them, when they leave [nearest town] to come to the school, they know that they are going to a certain place and because we will be expecting them, they become a visitor. They know that they are coming here to meet the children. That is why they come with the parcels, but the tourist is someone who probably will not visit a place carrying something to give. But for the visitors at times, they ask beforehand, before they come, 'What do you need? What are your needs?' Or at times, maybe the guide will ask, 'what are your needs?' Then we tell them.
>
> (Zibusiso)

For Zibusiso, the fact that the tourists ask prior to visiting what the school needs shows that they are not a tourist. By considering the needs of the local community and giving back, they become *more* than a tourist. The discursive positioning, that visitors are people who bring gifts, and tourists are people who do not bring anything to a destination, is a positioning through which traditional common discourses surrounding mass tourism can be seen. In the conventional sense of mass tourism, tourists are considered to bring nothing to a destination and may even leave a destination in worse condition that when they began (Harrison & Sharpley, 2017; Mowforth & Munt, 2009; Week, 2012). By identifying a tourist as a 'visitor' this enables the tourist to feel that they are not a tourist in this space, and they may believe that the tour is something unique to their visit, rather than something which happens up to three times a week.

Ayanda agreed with Zibusiso, they become visitors when they pre-plan and visit with the intention of providing gifts or contributions,

> They will be coming maybe to South Africa, Botswana, Zambia and Zimbabwe or most of the countries at one go. But to us, now especially, when they come – because I think they have liaised before – they know

that 'when we go to Zimbabwe, there is a special school which is ours. Where we do something, where we are going to go visit'. So, to us, these are our visitors. They are coming specially for us. They planned for us. So, to us, they can be tourists, yes and outside the school they are. But not to us, they are…. to us they are not called tourists, they are our visitors. So, we have learnt to possess them.

(Ayanda)

Ayanda provides more nuance to the term visitor. She identifies that whilst they visit multiple places on their holiday, they only become a visitor when they arrive at Matopo School. Ayanda refers to the school as *a special school which is ours*, suggesting that the school understands the need for the tourists to feel ownership and affinity with the school. Yet, Ayanda also describes how the school feels an ownership over the tourists saying, *we have learnt to possess them*, and *they are our visitors*. There is a discursive positioning which forms a relationship between the tourists and the school; in that, an ownership exists when the tourists visit the school, and they become *ours*. It draws the tourists into a donor-obligation nexus (Chouliaraki, 2010; Hernandez, 2018). In this nexus, the school recognises the subject positioning of becoming a 'donor' and provides opportunities for the tourist to take this up. However, the positioning both enables and constrains the conditions of possibility for the tourists and the school; in some ways it limits the ways a tourist can 'be' in the school. As outlined previously, the tourists are constituted as having love and affection, meaning their behaviour is governed by the donor-obligation nexus, and they may feel compelled to donate, even if they are not interested in supporting the school.

The tourists become visitors when they step inside the school grounds. But, according to the teachers, they are positioned in a more distinct manner when they interact with children at the school. Ayanda explains how the tour works, and reveals this naming of the tourists as 'friends':

We allow them when they come in, and if they have time, they can check the children's work whilst the kids are writing or even on our shelf. We allow them to open and see how the kids are going, what they are doing, what they are writing, even those who are grown up like the grade five, when I was teaching them, I would say 'you take your friend, and you show them your books'. They will go where they sit and say, 'this is how I work' and then their friends will comment and so forth, even take pictures of what they are seeing the kids do.

(Ayanda)

During the school tour, the tourists are provided with almost unlimited access to the classrooms, and to the children's schoolbooks. The access may contribute to the feelings of ownership as the tourists feel they can know what the school's values are, the way the children are working, and the work ethic of the teachers.

Ayanda makes use of the term friend, *take your friend* and *their friends will comment*. The term 'friend' is part of the discursive functioning of the power relationship. In an exploration of Cuban tourism Simoni (2014) states, 'friendship, much like love, calls for a certain commitment and continuity in [the] relationship' (p. 286). In the act of referring to tourists as 'friends' the school entangles the tourist in an act of mutual recognition about how a friend might act or be. The positioning of a friend is enacted, in part, through the idea of the gaze – through which tourists and the school engage in behaviour as part of their own regulation. Foucault (1975/1991) described this, 'he is who is subjected to a field of visibility, and who knows it... he inscribes in himself the power relation in which he simultaneously plays both roles; he becomes the principle of his own subjection' (pp. 202–203). Through technologies of power, such as gaze, individuals will self-govern to conform to the dominant discursive constructions of how a 'friend' may look or act.

The naming of the tourists as 'visitors' and 'friends' is an important element of the tourism. The naming allows the tourists to feel a level of ownership over the school, and it reminds the teachers and children they need to regard the tourists with a certain level of care. Further, the naming develops a sense of a deeper relationship, beyond the traditional relationships associated with mass tourism. Many tourists seek authenticity when they travel and hope to see the 'real' culture (Duval, 2004; Leonard, 2011; Lozanski, 2010a; Mkono, 2011; Week, 2012). By using the words visitor or friends, the school is working to create an image which constitutes tourists as special, unique, helpful, and distinctly *not* a tourist. This plays into the white saviour role, in which the white person has come to save the school – something that different tourists may choose to take up depending on their personality or previous experiences (Cravatte & Chabloz, 2008). The image works to allow the tourist to forget that they are involved in a mass tourism venture. By removing the word tourist, and replacing it with visitor or friend, it allows the tourist to forget that they may not be the only group to visit the school. Thus, it reinforces the illusion of the tour as a spontaneous spectacle, which happens unexpectedly for the group of tourists. These discourses of authenticity are well established in tourism ventures, and the school plays on the quest for authenticity.

The school founders also subscribed to this idea of the word *friend*, with Alison describing what she witnessed on the school tour,

> And this was, again, I only saw Adventure Tour groups, but the friends or whatever, the kids leading them around would come into the bead shop.
>
> (Alison)

In a similar manner, a teacher described the interaction with the tourists:

> When they come, first, we go to welcome them, from their car with our kids. Then we give them to their friends, I mean our kids.
>
> (Bridget)

For both Alison and Bridget, there seems to be a recognition that the word 'friend' is not a truthful description of the tourists. In both examples, they use the word friend and then a second phrase to show their caution for using the word. For Alison, she follows the word friend with *or whatever*, and Bridget immediately corrects herself, saying *I mean our kids*. Tourism scholars refer to the 'front' and 'back' stage of tourism, through which certain actions are positioned (Edensor, 2001; Gibson, 2009). The backstage is a discursive construction which identifies all actions that occur away from tourist gaze. The backstage usually refers to unconscious behaviours such as language choice, dress, and behaviour. In other cases, 'backstage' occurs in the deliberate actions which are taken to maintain the image of the tourism. For example, an explicit action conducted in the backstage to maintain the image in the frontstage would be the story told in the previous chapter about the children who had fought in front of tourists. In this case, both Alison and Bridget may have realised they were using 'frontstage' language in a backstage setting and adjusted their manner of speaking to reflect that they were not talking to a tourist. At some level, it seems that the teachers and tourism personnel know they are enwrapped in a need to portray the 'authentic' school and to minimise the image of Matopo School as a mass tourism destination.

The language of 'friend' and 'visitor' discursively constitutes the tourists into a relationship with the school. This relationship and dominant discourses of care, authenticity, and obligation frame the power relations in the school and, in turn, the conditions of possibility for action by both the tourists and the teachers. This is an important factor that highlights how the tourism at Matopo School is not necessarily a one-way gaze of tourist upon child; indeed, the school community gazes back and adjusts their behaviour to maximise tourist experience to maximise potential philanthropic funds.

## Summary

I began by outlining the poor-but-happy discourse and the ways in which the children at the school are constituted by these discourses of development and white saviourism. The teachers, in their role as directors in the movie metaphor, work to produce the image of the poor-but-happy child as they know this is a significant reason behind the funding of their school. I have also outlined the affective state that the tourists appear to be in when they visit the school, the affective state of love and care for the children. This is an emotive state, in which there is an active encouragement for tourists to 'feel'. As noted by the teachers, this love does not extend to them. There are taken-for-granted assumptions which are embedded in these declarations of 'love' or care, in which there are mutual benefits of being the receiver/gifter. In line with this, the naming of the tourists is important, as it positions the tourists in a different type of relation: that of a visitor or a friend. By inscribing notions of care and friendliness associated with these two subjectivities, the children in the school are expected to act with a certain level of hospitality towards the tourists.

## References

Aitchson, C. (2001). Theorizing Other discourses of tourism, gender and culture: Can the subaltern speak (in Tourism)? *Tourist Studies, 1*(2), 133–147.

Aitken, S. C. (2001). Global crises of childhood: Rights, justice and the unchildlike child. *Area, 33*(2), 119–127.

Bandyopadhyay, R. (2019). Volunteer tourism and "The White Man's Burden": Globalization of suffering, white savior complex, religion and modernity. *Journal of Sustainable Tourism, 27*(3), 327–343. https://doi.org/10.1080/09669582.2019.1578361

Bandyopadhyay, R., & Patil, V. (2017). 'The white woman's burden' - the radicalized gendered politics of volunteer tourism. *Tourism Geographies, 19*(4), 644–657. https://doi.org/10.1080/14616688/2017.1298150

Bell, K. M. (2013). Raising Africa? Celebrity and the rhetoric of the white saviour. *PORTAL Journal of Multidisciplinary International Studies, 10*(1). https://doi.org/10.5130/portal.v10i1.3185

Brinkman, S. (2018). The interview. In N. K. Denzin & Y. S. Lincoln (Eds.), *The SAGE Handbook of Qualitative Research* (5th ed., pp. 576–599). SAGE Publications Ltd.

Burgold, J., & Rolfes, M. (2013). Of voyeuristic safari tours and responsible tourism with educational value: Observing moral communication in slum and township tourism in Cape Town and Mumbai. *DIE ERDE - Journal of the Geographical Society of Berlin, 144*(2), 161–174. https://doi.org/10.12854/erde-144-12

Carpenter, K. (2015). Childhood studies and orphanage tourism in Cambodia. *Annals of Tourism Research, 55*, 15–27. https://doi.org/10.1016/j.annals.2015.08.010

Chouliaraki, L. (2010). Post-humanitarianism Humanitarian communication beyond a politics of pity. *International Journal of Cultural Studies, 13*(2), 107–126. https://doi.org/10.1177/1367877909356720

Cohen, S. A., & Cohen, E. (2017). New directions in the sociology of tourism. *Current Issues in Tourism, 22*(2), 153–172. https://doi.org/10.1080/13683500.2017.1347151

Cravatte, C., & Chabloz, N. (2008). Enchantment and solidarity. *Tourist Studies, 8*(2), 231–247. https://doi.org/10.1177/1468797608099250

Crossley, É. (2012). Poor but happy: Volunteer tourists' encounters with poverty. *Tourism Geographies, 14*(2), 235–253. https://doi.org/10.1080/14616688.2011.611165

Duval, D. T. (2004). Cultural tourism in postcolonial environments: Negotiating histories, ethnicities and authenticities in St Vincent, Eastern Caribbean. In M. C. Hall & H. Tucker (Eds.), *Tourism and postcolonialism: Contested discourses, identities and representations*. Routledge.

Edensor, T. (2001). Performing tourism, staging tourism: (Re)producing tourist space and practice. *Tourist Studies, 1*(1), 59–81.

Fanon, F. (1986). *Black skin, white masks*. Pluto Press.

Foucault, M. (1975/1991). *Surveiller et punir: Naissance de la prison* [Discipline and Punish: The Birth of the Prison] (A. Sheridan, Trans.). Penguin Books Australia Ltd.

Foucault, M. (1976/2008). *La Volonté de-savoir* [The History of sexuality: Volume 1] (R. Hurley, Trans.). Penguin Group (Australia).

Foucault, M. (1980/1988). Power, moral values and the intellectual (interview with Michael Bess). *History of the Present, 4*(1–2), 11–13.

Frazer, R., & Waitt, G. (2016). Pain, politics and volunteering in tourism studies. *Annals of Tourism Research, 57*, 176–189. https://doi.org/10.1016/j.annals.2016.01.001

Gibson, C. (2009). Geographies of tourism: (un)ethical encounters. *Progress in Human Geography, 34*(4), 521–527. https://doi.org/10.1177/0309132509348688

Harrison, D., & Sharpley, R. (2017). Introduction: Mass tourism in a small world. In D. Harrison & R. Sharpley (Eds.), *Mass tourism in a small world* (pp. 1–14). CPI Group (UK) Ltd. https://doi.org/10.1079/978178068545.0000

Hernandez, K. (2018). Can the subaltern be seen? Photographic colonialism in service learning. *Qualitative Research Journal*, *18*(2), 190–197. https://doi.org/10.1108/QRJ-D-17-00051

Judge, R. C. (2015). Emotion, volunteer-tourism and marginalised youth. In M. Blazek (Ed.), *Children's emotions in policy and practice* (pp. 157–173). Palgrave Macmillan.

Judge, R. C. (2017). Class and global citizenship: Perspectives from non-elite young people's participation in volunteer tourism. *Tourism Recreation Research*, *42*(2), 164–175. https://doi.org/10.1080/02508281.2017.1303894

Kesby, M., Gwanzura-Ottemoller, F., & Chizororo, M. (2006). Theorisingother, 'other childhoods': Issues emerging from work on HIV in urban and rural Zimbabwe. *Children's Geographies*, *4*(2), 185–202. https://doi.org/10.1080/14733280600807039

Leonard, M. (2011). A tale of two cities: 'Authentic' tourism in Belfast. *Irish Journal of Sociology*, *19*(2), 111–126.

Liebel, M. (2019). Colonialism and the colonisation of childhoods in the light of post-colonial theory. In T. Kleibl, R. Lulz, N. Noyoo, B. Bunk, A. Dittmann, & B. Seepamore (Eds.), *The Routledge Handbook of Postcolonial Social Work*. Routledge.

Lozanski, K. (2010a). Defining 'real India': Representations of authenticity in independent travel. *Social Identities*, *16*(6), 741–762. https://doi.org/10.1080/13504630.2010.524781

Lozanski, K. (2010b). Independent travel: Colonialism, liberalism and the self. *Critical Sociology*, *37*(4), 465–482. https://doi.org/10.1177/0896920510379443

McClintock, A. (1995). *Imperial leather: Race, gender and sexuality in the colonial contest*. Routledge, Inc.

Mhazo, T., & Thebe, V. (2020). 'Hustling Out of Unemployment': Livelihood responses of unemployed young graduates in the City of Bulawayo, Zimbabwe. *Journal of Asian and African Studies*, *56*(3), 628–642. https://doi.org/10.1177/0021909620937035

Mills, C., & Lefrançois, B. A. (2018). Child As Metaphor: Colonialism, Psy-Governance, and Epistemicide. *World Futures*, *74*(7–8), 503–524. https://doi.org/10.1080/02604027.2018.1485438

Minca, C. (2010). The Island: Work, tourism and the biopolitical. *Tourist Studies*, *9*(2), 88–108. https://doi.org/10.1177/1468797609360599

Mkono, M. (2011). The Othering of food in touristic eatertainment: A netnography. *Tourist Studies*, *11*(3), 253–270. https://doi.org/10.1177/1468797611431502

Mostafanezhad, M. (2013a). The geography of compassion in volunteer tourism. *Tourism Geographies*, *15*(2), 318–337. https://doi.org/10.1080/14616688.2012.675579

Mostafanezhad, M. (2013b). 'Getting in touch with your inner Angelina': Celebrity humanitarianism and the cultural politics of gendered generosity in volunteer tourism. *Third World Quarterly*, *34*(3), 485–499. https://doi.org/10.1080/01436597.2013.785343

Mowforth, M., & Munt, I. (2009). *Tourism and sustainability: Development globalisation and new tourism in the Third World* (3rd ed.). Routledge.

Sheridan, V., Landy, D., & Stout, V. (2019). The return of the 'Black Babies': How development education affects schoolchildren's attitudes to the majority world - the Trócaire 2012 Lenten campaign. *Race, Ethnicity and Education*, *22*(6), 858–874. https://doi.org/10.1080/1361332.2017.1417829

Simoni, V. (2014). From tourist to person: The value of intimacy in touristic Cuba. *Journal of Tourism and Cultural Change*, *12*(3), 280–292. https://doi.org/10.1080/14766825.2014.934054

Simpson, K. (2004). 'Doing development': The gap year, volunteer-tourists and a popular practice of development. *Journal of International Development*, *16*(5), 681–692. https://doi.org/10.1002/jid.1120

Smithers, K. (2022). 'What is your name, where do you come from, what is your grade?' Using art-based interviews to highlight the experience of children hosting school tours in Matabeleland North, Zimbabwe. *Tourism Recreation Research*, *48*(6), 912–924. https://doi.org/10.1080/02508281.2022.2133812

Staples, A. J. (2006). Safari adventure: Forgotten cinematic journeys in Africa. *Film History: An International Journal, 18*(4), 392–611.

Toomey, N. (2017). Humanitarians of Tinder: Constructing whiteness and consuming the Other. *Critical Ethnic Studies, 3*(2), 151–172.

Wale, K., & Foster, D. (2007). Investing in discourses of poverty and development: How white wealthy South Africans mobilise meaning to maintain privilege. *South African Review of Sociology, 38*(1), 45–69. https://doi.org/10.1080/21528586.2007.10419166

Week, L. (2012). I am not a tourist: Aims and implications of "travelling". *Tourist Studies, 12*(2), 186–203. https://doi.org/10.1177/1468797612454627

ZIMSTAT. (2017). *Education report 2017.* ZIMSTAT. http://www.zimstat.co.zw/wp-content/uploads/publications/Social/Education/Education-Report-2017.pdf

# 7    Breaking the fourth wall

## The white saviour as protagonist

The protagonist is the character who undergoes some sort of transformation or journey throughout the film. They are not always the main character, but usually play a large role in the film. Whilst the protagonist in this case is not the star, they are the one who have undergone a 'transformation' by visiting 'Africa'. This transformation is one of the discourses, associated mainly with volunteer tourism, that circulate about the life-changing aspect of travel and interaction with the marked Other (Crossley, 2012; Mostafanezhad, 2013c).

As if they are following the movie script, the white saviour (tourist) will respond to the perceived needs of the poor-but-happy children in accordance with the movie script, rather than asking what the school might need. According to the research literature, the white saviour complex is in part achieved through an unfamiliarity to the needs of the people they hope to serve. Instead, the white saviour's perception of needs may be based in their own understandings – which means they will always be satisfied about a job 'well-done' (Bandyopadhyay, 2019; Bell, 2013; Bex & Craps, 2016; Cammarota, 2011; Goudge, 2003; Hughey, 2010; Lough, 2015; Lozanski, 2010). The white saviour narrative is one which individualises foreign aid and development, it posits that one person can make a difference in the lives of the marked Other (Bell, 2013). Further, the white saviour discourse facilitates the transference of gaze, and in my field notes there are accounts of frequent photography by tourists who appeared to have little regard for gaining consent from adults or children. Returning to the film metaphor, the white saviour as the protagonist focuses solely on their own charity and feelings of satisfaction; that is, the focus is on their character development alone.

In the previous two chapters I examined the directors/producers of the film and the starring role: the poor-but-happy child. I now explore how the teachers, tourism personnel, and the children are always-already positioned by many stereotypes and discourses as part of their positioning in the global hierarchy, which continues to privilege Minority World ways of knowing and doing. I begin by exploring the problematic white saviour role in depth, particularly in relation to gift-giving practices the tourists enacted whilst in the school. Second, I explore the role of whiteness and the white saviour role in Development more broadly through a discussion of one tension involved, that

DOI: 10.4324/9781003407980-7

of having external funders financing the school. Last, I explore whiteness in Zimbabwe through a discussion of the relationship of the local community to the school founders.

### The protagonist: the white saviour

I now detail the ways in which the tourists enacted their own visions of a 'white saviour' and the ways that this links to the larger web of power which constitutes the 'white saviour'. As established in the previous chapter, the teachers often felt there was a focus solely on the children, in terms of gift-giving. Further, as previously established, the teachers manage the disruptions from the tourism because they believe the benefits of the tourism outweigh the pedagogical implications of the disruptions. To begin, I focus largely on the gift-giving practices of the tourists and the ways the tourists enacted their visions of white saviour in the school.

Discourses regarding the white saviour circulate through media reports of celebrity humanitarianism, for example through reports of celebrities like Madonna, Bono, and Angelina Jolie visiting foreign countries to provide aid (Bandyopadhyay & Patil, 2017; Bell, 2013; Mostafanezhad, 2013a). Through these images of celebrity humanitarianism, Minority World tourists can imagine and prepare to enact 'vivid scripts for their own fantasy roles in the continent's salvation' (Garland, 2012, p. 7). Often, local people in the movie (real or imagined) are not allocated a speaking role, instead are used as supporting characters and as part of the set design which enhances a heterotopia based in the image of Africa (Salazar, 2009).

The demographic of the tourists was mostly white American, usually aged over 50 and semi-retired, as this was the main clientele base of the companies who visited the school. During the fieldwork, these were the only tourists that I encountered. However, this is not to say that black African American tourists did not visit the school, they most certainly did on occasion. It could also be argued that for the school, and the tourism industry more broadly, the marker of 'American' is significantly more important than the marker of 'white'. Thompson, the black school founder, details his past experience with this:

> I was doing intern training in the camp, as a senior guide/manager. And I basically said to all the guides, I said, 'please, go to the list of all the guests coming into the camp. I would like you to pick all the guests that you think are going to be good and leave me the people you don't want'. So, of course they pick all the Americans and everybody and left the British couple.

As will be detailed later in this chapter, Thompson had reflected considerably on the role of whiteness in shaping his business, his life, and his charitable endeavours. In this extract Thompson identifies that in tourism, being 'American' is often a marker of wealth. In contrast, Judge (2016) explored the

experiences of young, black, British volunteer tourists in Zimbabwe and identified 'there are relations that establish the objects of charity as black African
bodies and the resource-bearing donors as white Western bodies' (p. 245).
These relations may have played out in similar ways at Matopo School, but
they were not experienced during the time I was there.

To return to the metaphor, the white saviour is always the protagonist of the
movie (Bell, 2013). In an example of this, I will now consider the description
given by Greg, a white tour guide for Education Tours, regarding the practice
of visiting the market prior to the school. The market visit allows tourists to
haggle and buy food as a gift for the school. Greg identifies that the market
visit allows the tourists to play two roles,

> They can do all that side of it and then they are giving it to the school as
> well, so that is… I try to encourage that more than actually giving money.

One role performed by the tourists is the role of the 'haggler' in the marketplace and the second role is the role of altruistic tourist who provides suste-
nance to the school, as the white saviour. The school has a well-established
feeding programme which aims to provide nutrition to the school children. The
feeding programme is paid for by school fees, and the children are provided
with two hot meals a day. Whilst the extra food provided by these tour groups
was very welcome, the school has pre-existing supply chains, and procedures,
for their feeding programme.

The giving of gifts is complicated by the policies of tourism companies. For
example some companies may implement some form of a briefing through
which individual gifts are discouraged and a group financial donation is
encouraged (Cravatte & Chabloz, 2008); however, financial donations are
sometimes discouraged by tour guides, such as Greg, and the teachers feel that
the tourists do not trust them with financial donations. These complexities
might make it difficult for a tourist to determine whether their donation is one
of use, as the discursive position in tourism advertising is the white saviours'
donations are *always* useful.

In contrast to the practices of Greg, Ayanda, a senior teacher, describes a
monetary donation provided to the school,

> One of the visitors found out that our diet was lacking meat because of the
> expenses we could not afford. So, they volunteered to say I will throw in a
> lump sum and you can buy some meat for the school.
>
> (Ayanda)

By providing money to the school, rather than perishable food items such as
tomatoes or cabbages, the school could purchase meat for meals rather than
having to store items that they may already have in stock. In Chapter 5, I discussed the ways in which Greg is constrained by his positioning within the
network of relationships, and the ways he is governed by how a tour of 'Africa'

should look. Greg's decision to take the tourists to the market to buy vegetables is reflective of the dominant discourses which constitute a heterotopic image of Africa as needy and impoverished, but is also shaped by the white saviours' desire to 'giveback' to destinations they visit.

The story by Ayanda demonstrates how the role of white saviour is fraught with tensions. In the story by Ayanda, the tourist listened to the needs of the school and acted upon these needs. Discourse is a 'multiplicity of discursive elements that come into play in various strategies' (p. 100) and 'there can exist different and even contradictory discourses within the same strategy' (p. 102) (Foucault, 1976/2008, p. 102). Although I have addressed the tourists as 'the white saviour', the white saviour discourse is unstable and exists in different forms. However, as part of delineating the white saviour 'discourse' I focus on those white saviours who do not consider the needs of the people they aim to 'save'. For example, as shown by Ayanda, donations are useful, if the context and needs of the community are taken into account. It is within the instability and complexities of the white saviour discourse that the power relations are embedded.

When asked about how the tourist gifts are organised and managed in the school, Astrid, the teacher involved with organising and distributing the gifts, said the following:

> I have seen what it does to these kids. In our culture when we are, like in the primary school age, you are not supposed to ask anything from an adult. We are taught, like an elder person has to see on her or his own that you are in need. Then he will help you. In whatever way he thinks is right. But in this case, our kids they are now begging... They are not supposed to say, 'I do not have this'. Because what we say in our culture, if you are an adult or parent or elder, you are a parent to the society, to the whole community. If you see a child who is troubled, you help that child. Without having so much knowledge of the child, without that child begging to you, saying 'can you help me with this?'
>
> (Astrid)

Astrid identified that the tourists' gifts disrupted the cultural expectations and norms of the community regarding asking for support from adults. Later in her interview, Astrid said, 'we are so desperate for the funds that they are giving' and outlined the ways in which the school often ignored things it did not like, in order to cater for the tourists. Astrid demonstrates the ways in which the need of the school creates a complex relationship of power in which the receiver of the donation is drawn into a nexus of reciprocity and obligations to the (relatively) wealthy donor (Chouliaraki, 2010; Hernandez, 2018). These obligations might include being appreciative of the donation and ensuring the donor *feels* duly appreciated by the recipient. The donor-recipient relationship is guided by discursive notions of development; that is, they are guided by what it means to be/become a donor and a recipient. In this way, the donor/recipient relationship creates conditions of possibility through which the teachers and children are governed.

Jackson (2013) details how the white saviour may have very little regard for the needs of the local community, rather they enforce their own Minority World ideals when providing what they perceive addresses the 'need' of the community. The privileging of Minority World ideals is reflective of the white saviour discourse which constitutes Minority World intervention as efficient, modern, and correct (Bell, 2013). The idea that the white tourists can know what is best for the local people is an example of the privilege of whiteness at work. In an example of this, the tourists utilise 'pack for a purpose' to fulfil their role of white saviour.

'Pack for a purpose' (https://www.packforapurpose.org) is an example of company policies that complicate the *possible* roles and actions of the white saviour. It involves packing small items within luggage, to give to those in the destination who are in 'need'. The companies who provided tours at Matopo School included information about pack for a purpose in their pre-departure information and it is strongly encouraged. Most tour groups who visited the school arrived with pencils and other small gifts for the school. Tourists usually presented the gifts when getting off the tour bus, or during the school history component. These gifts were things the school could use and represented a very minor role in the interaction with the tourists. It could be argued that pack for a purpose is an example of the ways in which tourist behaviour is constrained. In this case, tourists who visit Matopo School on a tour are constituted, prior to departure, as a white saviour through the pack for a purpose initiative. Tourists may feel compelled to bring a donation or act in other ways, as this is part of the discourses of becoming a tourist in sub-Saharan Africa. Whilst some engaged in small gift-giving, there were examples of donations that suggested a larger investment in 'saving' Africa.

One 'pack-for-a-purpose' gift exchange highlights the cultural complexities of gift-giving. A male tourist had donated enough reusable menstrual pads for the senior class of Matopo School. During the tour, the girls in the senior class and the female teachers had been called to a classroom to be provided with instructions on how to use reusable pads. The presentation, given by an older local female, was arranged by the charitable organisation associated with Adventure Company and was intended to be culturally sensitive. There was, however, a sense of discomfort during the presentation because the donor was also present. The donor was a middle-aged white male who took pictures of the children and the presenter during the talk. An extract from the field notes explores this further,

First, Samantha told them to thank him, his name was Thomas. They said 'Thank you Thomas' half-heartedly. I think because it was the start of the presentation and they had not received the pads; they did not fully understand who/why they were saying thank you to. Samantha told them to say it again because it was not proper, then he asked it to be repeated for a third time so he could film it! After the presentation about the usage of menstrual pads, there was lots of photos in different combinations such as with the girls with the pads, the man with the people who had presented, the man with the kids showing him around the school etc. It would have been 12:15pm by the time they left so the kids were interrupted, after the initial proceedings, from 9:30am–12:15pm. And the Adventure Company visit is meant to be less of an interruption!

There are at least two matters to explore in this gift exchange that are relevant for my discussion of the tensions inherent in gift-giving. First, the male performed a culturally insensitive act by being present in a discussion that is usually for women only. Second, he asked for the 'thank you' to be repeated so he could film. In Zimbabwe, discussion of menstrual cycles is something considered to be private, and generally only discussed in the presence of females, usually with a maternal aunt (Power et al., 2004). During the presentation there was a visible discomfort among the children and the female teachers.

The second issue is around the acknowledgement of the gift. There are unequal power relations present in the interactions in the school, operated through the discourses of development, in which the recipients are bound by obligations to the donors (Chouliaraki, 2010). The discourses associated with the donor/recipient relationship govern the interaction and the recipient must be grateful for the contribution by the donor, no matter how big or small or whether it is something they needed. For the donor, their success is measured through the tone and quality of the recipients' 'thank you' (Hernandez, 2018). In a Foucauldian understanding, conditions of possibility are intimately tied to discourse; Foucault (1976/2008) asked, 'what were the most immediate, the most local power relations at work? How did they make possible these kinds of discourses, and conversely, how were these discourses used to support power relations?' (p. 97). Discourses and power relations work to create the conditions through which subjectivities are constituted and through which regimes of truth form what can be 'known'. In this scenario, the conditions of possibility ensure that the school complies with the wishes of the donor and with the requests of the third-party charitable organisation who had organised the presentation. Whilst the school *could* say no to the culturally insensitive requests of the donor, it would involve a level of risk – they are positioned unequally in the power relations and refusal might mean a discontinuation of funding.

Perhaps one of the most problematic aspects of the white saviour complex is the demand for recognition of the 'good' deed. In the extract, Samantha is a black Zimbabwean woman who works for the charitable organisation funded by Adventure Company and she governs the girls by asking them to repeat their 'thank you' in a more correct manner. Power is located, 'not so much in a person as in a certain concerted distribution of bodies, surfaces, lights, gazes; in an arrangement whose internal mechanisms produce the relation in which individuals are caught up' (Foucault, 1975/1991, p. 202). It is within power relations that individuals are caught, in the web-like networks of relationships which work to govern. Samantha is 'caught up' (Foucault, 1975/1991, p. 202) in the 'panopticisms of every day' (p. 223) and in this case the girls must conform to the dominant discourses of what a 'good' recipient looks like, with their behaviour governed by Samantha, who asks them to repeat the 'thank you' until they have met an appropriate level of appreciation. Further, Samantha governs the girls under the gaze of the white man who is sitting in the room.

In the scenario described, the white saviour may only perform their acts of 'saving' if they have recipients for their good deeds and are then recognised in their home country. The donor-recipient imbalance inflates the sense of importance of the donor and compels the donor to act as if the recipients should always be gracious and appreciative of any acts of kindness (Arnett, 2016; Frazer & Waitt, 2016). The regime of truth continues the global hegemony of Minority World ways of knowing by ensuring that the recipients do not tell the donor what they might really need. Further, it confirms a pervasive regime of truth in Development – that white people know best and know what local communities need (Escobar, 1984; Jackson, 2013; Sheridan et al., 2019). A regime of truth does not represent the 'truth' but rather, it represents the way things become 'true' through relationships of power and claims to knowledge. Regimes of truth are tied to subjectivation, 'since every regime of truth asks to the individuals who are implicated in it, a specific self-constitution' (Lorenzini, 2015, p. 5). The white saviour metaphor works to legitimise the regime of truth; there is no need for the saviour to ask whether they are helping, the proof is in the 'thank you' they receive.

Recognition is usually depicted through the circulation of photographs of the good deed. Hernandez (2018, p. 193) refers to these photographs as 'helpies', in which photographs are used as a method of gaining recognition usually through social media comments. These 'helpies', selfies through acts of service, are distributed through social media and provide a way to receive recognition for acts of service. The children, teachers, and presenter are placed within conditions of possibility through governing discourses of donor-recipient obligations whereby they have very little choice but to participate in the photograph/ video. As Hernandez (2018) explains, in the Majority World gift exchanges involving wealthy donors include 'tacit exchange expectations that accepting the gift automatically extends photo rights to the giver' (p. 193). To resist these expectations would involve a level risk for the school. For example, if the school does not comply Adventure Company may withdraw tour groups from the school or provide less support and donations.

The relationship between donor and receiver is governed by discursive notions of Development: all participants in the situation are guided by what it means to be/become a donor and a recipient. In this way, the donor/recipient relationship functions as the model through which the norms for behaviour at the school are governed. The act of filming the 'thank you' from the children renders them as passive receivers who are expected to be subservient in an exchange where rights of a photograph are automatic and anticipated. The discursive construction of the relationship between donor and recipient informs this exchange and ensures that the subaltern, who may be able to speak through the act of thank you, is never properly heard. Further, the filming of the 'thank you' recentres the white saviour in the narrative, which is then distributed for Minority World consumption.

The construction of the white saviour has a dual function in which the tourist perceives oneself as a saviour, but the school understands this subject position and takes it up, allowing the tourist to enact their saviour role. In the following extract from my field notes there is an emotive sense of frustration, as I grappled with my own uncertainty about my interactions with the tourists and the role I should play in their visits. An example of the opportunity for tourists to enact the white saviour subject position is from the field notes when two tourists address the Grade 5 class:

> I am sitting in the Grade 5 classroom, helping to sew some bags and Ayanda asks the visitors to inspire the children. The first lady who stands up says something that was neither here nor there in terms of inspiration but did condescendingly ask the children, 'do you know what a nurse is?' Sorry, but that is a basic thing.
>
> The next guy who stands up is an electrical engineer and he tells them if they want to be an engineer, they will have to learn lots of maths, and then he turns to us teachers and says, 'Do you teach math here?'

It might seem odd that the school would ask tourists to inspire the children. However, at the school, there was a belief that most of the people who visited the school were retired school teachers – after all why else would someone have an interest in a school? This belief was probably furthered by the tourists, who would sit with children and begin helping them with their work or begin teaching the class (see also Shelab, 2011, p. 284). Through Ayanda asking the tourists to inspire the class, there is a recognition by the school of the role that the tourists hope to play in the scenario: potentially the role of the white saviour or altruistic teacher. In the discursive construction of the saviour, a tourist can only become a saviour if they have willing people to enact the saving upon. In this example, Ayanda is recognising this subject position and allowing the tourists an opportunity to play the role. In this act of recognition, the saviours begin their story by assuming a deficit of knowledge about Minority World ways of knowing: as the saviour is there to inform and impart knowledge on those who are less than (Harrison, 2015; Kapoor, 2004; Kothari, 2006).

Now I turn to examine the underlying assumptions in the questions, *do you teach math here?* and *do you know what a nurse is?*. These assumptions reflect the broader discourses that constitute certain people as white saviour and others as those in need of saving. They are reflective of the 'deficiency model' of Development whereby countries in the Majority World are measured on their 'lack' of Minority World knowledges (Jackson, 2013; Kothari, 2006). First, the question by the woman: *do you know what a nurse is?* This question is thoughtless, as immediately prior to the school tour during the 'school history' component of the tour the tourists had been shown the new medical clinic which was being built for the village. Further, the question suggests that the children will not know what a nurse is, as basic healthcare and medicine would not be available in the rural area, due to the 'primitive' setting of the African continent. This builds the image of Zimbabwe as an unexplored frontier in which there is a lack of civilisation and Minority World amenities. In this unexplored frontier, the white saviour is constituted as having a role to play in educating the local people in Minority World ways of knowing and doing (Briggs & Sharp, 2006; Wale & Foster, 2007).

The assumption by the man who asks *do you teach maths here* shows his naivety and perhaps a lack of attention to the school history provided to him in the moments that had proceeded the question. This is not to suggest that the tourists should thoroughly research all their destinations prior to arrival, but it shows a significant oversight in terms of knowing the country rather than relying on stereotypes. For example, the company he visited with provides a detailed reading list for tourists to read prior to their journey. The list includes several books aimed at promoting an image of Africa which is not of just poverty and danger. Despite this preparatory reading list, the tourist draws on his past associations with the continent of Africa – that of uncivilised and uneducated – to assume that mathematics is not taught in the school.

There are various assumptions to unpack. Perhaps the man assumed that not all schools would be mandated to teach mathematics? His questioning is reflective of discourses of Development, in which countries in the Majority World can only develop through Minority World frameworks and ways of knowing (Bhambra, 2013). Further, it is reflective of the deficiency model of Development, as identified above. To give both of these tourists some allowance, it should be noted that not all adults are skilled, or practiced, in talking to children. There is a possibility the tourists may have said the first thing that came to their minds, rather than these questions reflecting their assumptions about the Zimbabwean school system. Most adults when put on the spot to talk in front of a group of children may grasp at the first thing that comes to their mind. It is in the response that the taken-for-granted assumptions and stereotypes come to light, in the 'implicit characteristic of the great anonymous, almost unspoken strategies which coordinate loquacious tactics' (Foucault, 1976/2008, p. 95). As Foucault argued, power relations are embedded in the social nexus, and not something which is outside of or separate from society (Foucault, 1982/2002). The taken-for-granted assumptions and stereotypes govern the tourists' interactions in the school as the discourses of Development are incredibly powerful and embedded in Minority World ways of knowing and doing.

**Complexities of the white saviour in action**

Within Development, race and class are often unspoken with an inherent assumption that all development workers are white and middle class (Escobar, 1995/2012; Goudge, 2003; Mowforth & Munt, 2009). Many argue that

Development, and work within it, is embedded with knowledge that privileges the Minority World (Faria & Mollett, 2014; Goudge, 2003; Mowforth & Munt, 2009). Kapoor (2004) identifies,

> Working in development inevitably positions us within a 'development discourse', where the North's superiority over the South is taken for granted, and Western-style development is the norm. Our encounters with, and representations of, our 'subjects' are therefore coded or framed in terms of an us/them dichotomy in which 'we' aid/develop/civilise/ empower 'them'.
>
> (p. 628)

I began this chapter by exploring the privileging of whiteness, the donor-obligation nexus and how these are ever-present in the gift-giving practices of the white saviour. I now continue the interrogation of the dominant discursive positioning of whiteness in Development which constitutes Minority World ways of knowing and doing as correct. First, I outline how the tourists constituted me as a white saviour, and then I outline the white saviour discourse present more broadly in these forms of charity and the tensions in Development practices.

### The role I was assigned

Whilst the tourists were enacting their own scripts and visions of what a white saviour may look like, tourists were also assigning the white saviour role to me, as the only white woman in the school. During the school tour, tourists often engaged me in conversations in which they referred to me as 'brave' or used similar adjectives to describe the work they perceived me to be undertaking. These discussions were uncomfortable, as the way they described my role in the school were examples of the white saviour discourses at work. Further they reflect the Minority World discourses that constitute volunteering in a Majority World country as automatically both virtuous and productive (Jakubiak, 2016; Snee, 2013).

The following extract is an example of a conversation written in my field notes, after one tourist sought me out whilst on their school tour. A conversation of this nature usually occurred every visit and generally interrupted the class activities:

> While I am sitting in class a visitor comes up to me and says, 'so what program are you with?' and I respond, 'I came here by myself' and she says, 'what made you decide to come here?' I explain briefly and she replies, 'I admire you. I could not do it, but I admire you. I was a teacher for 30 years, so I know I could not do it'.

The question, which asks *what program are you with* assumes that whiteness is out of place in the school and consequently the only reason a young white woman would be in a Zimbabwean classroom is through a 'safe' pre-organised programme. In volunteer tourism studies, participants often associate countries in Africa with risk and danger (Ansell, 2008; Garland, 2012). As discussed in Chapter 5, the discourse of risk and danger forms part of a heterotopic image of the 'setting' for the white saviour to conduct their work. Garland (2012) explores the role between this image and the discourses surrounding the white saviour, 'advertising materials assure tourists that they will be welcomed with open arms by their hosts, who are depicted in simple, stereotypic terms as needy cultural others, generic poor people eager and grateful for the assistance of benevolent Westerners' (p. 7). Thus, upon encountering a white woman, the conditions of possibility in the situation enable the tourist to assess my place within an established dominant discourse – the role of white saviour. This dominant discourse permeates interactions between tourists and local people in Zimbabwe, and Africa more broadly. An assumption made by the tourist is based on a regime of truth, which constitutes a white person in the school as a volunteer tourist, or a tourist who is volunteering 'safely' through a prearranged programme. If the tourist had encountered me on the street of the nearest large town, I may have been considered another tourist, or perhaps a local tour guide on a break. The positioning as a volunteer tourist is specific to the school and has discursive practices which are associated with it, which includes adjectives such as 'brave'. In the act of a white woman being in a rural Zimbabwean classroom, there are limited discursive positionings to be located within.

The assertion of braveness and admiration were not unique in the experiences recounted in my field notes; many of the tourists used the words 'brave', 'admiration', and similar adjectives to describe the work they assumed a white woman would be doing in the school. When the tourist described above says, '*I admire you*', there is an uptake of the white saviour discourse in which the white person is immediately marked as a hero (Cammarota, 2011; Mostafanezhad, 2013b). The white saviour discourse is reflective of the broader discourses, associated with a regime of truth of Development, which position Minority World volunteers as participating in risky but worthy work (Ansell, 2008). These discourses can be harmful as they govern interactions which occur in the Majority World and contribute to the belief that the white tourist can enact certain behaviours in the Majority World without consequences. These behaviours might include photographing children without permission (Crossley, 2012) or volunteering in a setting for which they are unqualified (Bargeman et al., 2018; Bauer, 2017; Monshausen et al., 2015).

Another common assumption was that I had joined the Peace Corps, followed by the declaration that I was 'doing very good work'. The Peace Corps is an American organisation, founded in 1960, whose work Jackson (2013) argues is underpinned by deficiency models of Development. These declarations were often at the beginning of the school tour and did not accompany observations of teaching practice or classroom behaviours – so one could argue there was no basis for the judgement of my work. In an extract from the field notes I begun to question why they would make these comments:

When the visitors come into my classroom one asks me, 'so, is this the equivalent of the Peace Corps?' and I say, 'No I just decided to come on my own without an organisation'. I often do not tell the tourists that I am a PhD student – I would prefer not to talk to them at all. They never ask why I am at the school, just what program I came with. The visitor says to me, 'What you are doing here is good work' and another says to me 'what a fantastic program' – yet none have asked me why I am here. By 'fantastic program', I am unsure if they mean me teaching at the school, or the school and the work of Thompson and Alison in general.

During the visits I often felt frustrated that my white skin immediately marked me as somebody who was 'brave' and 'doing fantastic work'.[1] Being sought out by the tourists was a particularly jarring experience, as I had hoped to work in a manner that respectful; through my research I tried to position myself on equal footing with the teachers, and I tried my best to minimise disruptions to the school (see also Harper & Smithers, 2023). The continued positioning of me as a saviour was a constant reminder not only of my own whiteness but the very ideals I had hoped to avoid, or at least minimise. In this way, my young, white body immediately encoded a plethora of discourses and conditions of possibility within the power relations, prior to any interactions or words spoken (Suzuki, 2017, 2018). The tourists may have sought conversation with me as whiteness is out of place in the otherwise 'African' landscape; therefore, it was contrary to the dominant discourses of what they would encounter at an 'African' school. Or, on the other hand, they may have sought me out as I represented something familiar, in an otherwise unfamiliar place.

Whiteness remains an invisible, unquestioned, and normative assumption (Christian, 2019; Knaus, 2019; Steyn, 2001; Suzuki, 2018). The privileging of whiteness allows white people to ignore the way that race structures every facet of life. In a Zimbabwean rural school, white skin is a marker of out-of-placeness. The tourists may not have expected to witness a white Minority World woman teaching; rather, dominant discourses constrain the conditions of possibility in such a way that they may arrive with an expectation to see black Zimbabwean subjectivities at work and to experience their heterotopia based in an image of Africa in real life. More generally, this is part of the 'Othering' practices of mass tourism at work – people travel to encounter and examine the exotic 'other' (Lozanski, 2010; Toomey, 2017). To be 'other' is to be counter to the *white* normative ways of doing and thinking. When the tourists see a white woman teaching or in a classroom setting, they placed me within their known discourse: the white saviour complex.

In another incident from the field notes, during supervision of the entertainment portion of the tour a woman approached me. The conversation began with 'are you the only white person here?'. The conversation was noted in my field notes as the tourist's opening line was unique, she was the only tourist to explicitly acknowledge my whiteness. *She continued the conversation by saying 'You are doing good work here! Have these. You do not need to share them if you do not want to. Keep them as a special treat'. She then handed me a bag of mixed chocolates.* The idea of the chocolates as a special treat is an example of the ways in which the heterotopia of 'Africa' plays out. The dominant discourses of 'Africa' as devoid of the comforts of Minority World life (Achebe, 1998; Matiza & Oni, 2014; Salazar, 2009) constrains the possibilities for the tourist to consider that chocolates might be readily available to me, or others in the school. Not only do the dominant discourses encourage her to envisage this heterotopia of Africa, but her own experiences in Zimbabwe may only be that of the marketplace (as described previously by Greg, a tour guide), rather than the local supermarkets which sell chocolates and other goods. *She continued the conversation by saying that she had intended to give them to the children but decided that I needed them more.* The exchange suggests that whilst she felt the children were needy, perhaps the only white person in the school would need them more. The assumption constitutes the work of a white woman as hard, but necessary work, in an unforgiving landscape with very little amenities, a well-established trope in volunteer tourism (Bandyopadhyay, 2019; Bandyopadhyay & Patil, 2017; Crossley, 2012; Godfrey et al., 2015).

Despite my discomfort at the conversations I experienced, they are reflective of the dominant discourses usually associated with volunteer tourism (Bandyopadhyay, 2019; Bandyopadhyay & Patil, 2017; Judge, 2017; Wearing et al., 2018). These discourses govern expectations of how a young, white female would act in an African school. In the metaphor that frames my analytical approach, my role didn't quite 'fit' and therefore, the tourists placed me within the known role: the white saviour.

**The tension of external funders**

Across my discussions with the Matopo School community, there was a tension between the school founders and Adventure Company, mainly resulting from a disagreement about the building of a classroom. Mary, a white tourism personnel with a previous career in teaching, was responsible for distributing the funding provided by a charitable foundation associated with Adventure Company. Using this classroom as an exemplar, I explore the tension present in implementing funding through external organisations and donors.

In the interviews with Mary, Thompson, and Alison it became apparent there were tensions in their relationship. In her interview, for example, Mary commented on the way one of the other companies run their tours:

| | |
|---|---|
| *Kathleen:* | …because Education Tours has the village tour as well, some of the cash goes to the villagers who are hosting. For the cost of the mealie meal…. like they go and stir the pot and then they have all… |
| *Mary:* | **[interrupts]** It is all contrived. That is the other thing. I did not find it authentic. And you see *we* celebrate cultures and authenticity. We do not want to change and skew a…. Anyway, does not matter. That is fine. |

When pressed further, she revealed she had not witnessed the village tour:

*Kathleen:*   So, have you gone on the village tour that they do?
*Mary:*       No. I have not [**shakes head and makes noise**]. I probably need
              to, but again it is a time thing. But now if we are getting a bit
              more organised, I probably will have the time, to do that.

The assertion, *it is all contrived* and then *you see we celebrate cultures and authenticity* is an example of the proliferation of claims to authenticity in tourism. Huxley (2004) provides a definition of authenticity in tourism as 'routine aspects of mundane quotidian existence, reproduced on an everyday basis, and located outside of the tourist centers and infrastructure' (p. 43). In tourism studies there have been debates about whether tours are, or ever could be, 'authentic' (Cohen & Cohen, 2017). The question I would rather ask is: who has the authority and under which conditions are they authorised to 'authenticate'? Mary may feel authorised to speak as tourism worker who has experienced 'authentic' tours. On the other hand, her authority to speak is reflective of her whiteness, through which white people are usually constituted with the right to speak in all situations (Knaus, 2019; Kobayashi & Peake, 2000; Wale & Foster, 2007; Yancy, 2004). It could be argued that Mary, as a white woman from Harare, cannot make the decision as to whether the village tour is contrived. Further, the declaration of authenticity speaks to a search for the cultural Other that is embedded in the discourses of 'authenticity' in tourism.

The relationship between Adventure Company, Mary, Thompson, and Alison was complex, with the main tensions surrounding the implementation of funding. Prior to our interview, Mary and I had met several times and on my previous visits to Zimbabwe. Perhaps due to this prior relationship, during her interview Mary asked *me* to appeal to Alison:

But I do think, bring to Alison the awareness that, you know, this is what it is. And [**pause**] sort of how it goes. So that he… [**pause**] yeah because like for instance, he could have had two classrooms up now, if we had done those prefab. Because, to me, if you are needing classrooms badly, let us just get them up. You can then change them years down the line, now he is going to have two classrooms which are going to take us this year to build them. In fact, even longer. Next year.

The 'pre-fab' buildings that Mary refers to are cheaper, quicker, and easier to build than brick classrooms. Mary is governed by the dominant discourses of Development; this is evident in her use of terms such as *this is what it is* and *how it goes*. The charitable organisation that Mary works with is a multinational charity which has policies in place in terms of quotes, building costs, and processes for implementing funding. Mary is constrained by her positioning within the organisation. The 'pre-fab' buildings were favoured by the charitable organisation that Mary worked for, as they are built by a Zimbabwean company and provide a quick turn around on donor money. As explored in previous chapters, donors often want photographic evidence that the money has been well spent and there may be a pressure to quickly spend the donor money, take the photograph, and send it to the donor. Mary's own ability to act outside of the dominant discourses of development is constrained by her position within the organisation. In fact, as her organisation relies solely on donor funds, if the donors are not pleased, Mary may no longer have a job. Further, Mary does identify the classrooms are quick to build and would need to be changed *years down the line*. Her frustration at the situation may be reflective of the pressures she had received from the organisation to conform to quick fix turnarounds to please existing and potential donors. Despite the dominant discourses which work to constrain her ability to act, Mary did resist the expectations of a quick turnaround and organised classrooms that would take longer to build. At the time of writing, a more permanent classroom had been constructed.

Thompson details the tension about the classroom in his own interview, referring to another school who received the same classrooms Mary wanted to install:

Adventure Company offered those metal cabins, but that is a temporary structure. It is hot, and low roofs, and small windows. I always think, make it different, what would make Matopo School different to other schools? There are the important things, health and good quality education, good infrastructure. If you do not do that, it is going to be like any deteriorating government school, and to me it is a waste of time. It is a waste of time.

In the field notes there were various accounts from the teachers at the school who detailed how there was an unusable classroom which had been one of the first buildings built at the school. The building, erected in late 2013, was in a state of disrepair by 2019 as it had been built from temporary building materials by a group who had travelled to the school to build it. The deteriorating classroom served as a visual reminder to the teachers and school founders about the concern of using temporary structures for classrooms. Favouring building quick temporary classrooms, rather than permanent structures, reflects Development trends more broadly which provide comfort to the white saviours who fund the project but in the long run provide very little infrastructure to the community (Jackson, 2013).

The differing perspectives on whether to build temporary classrooms is reflective of a larger tension in developmentourism: who has the right to speak about how donor money is spent, and how quickly? While I have only outlined one issue, namely that of a classroom, this tension is reflective of broader issues at work, such as the need for donor money to be spent quickly, with evidence of that spending.

## The white saviour and the local community

In keeping with the theme of the chapter, the white saviour, I now examine whiteness more generally in the local community. I have included this section here as the local community plays a role in the implementation of the tourism. The nearest regional centre to Matopo School is a tourist town. The main street is dominated by curio shops, safari company offices, and cafes. The town was often described to me during the fieldwork as a 'bubble' which exists separately from the rest of Matabeleland North. It is predominately set up for tourist consumption, with cafes owned and operated by multinational safari companies. The white tour guides and white people who worked for safari companies mostly live in a low-density housing area close to the main street. The teachers at Matopo School live in a high-density housing area which is much further from this main street. Near to the main street is a private primary school, whose fees are unobtainable for most, and are over US$1000/term, which is close to ten times that of Matopo School. Both Mary and Greg live in the low-density housing area.

First, I will situate both Greg and Mary in their role. Greg, a local white tour guide, is one of the 'producers' in the movie metaphor, who works to ensure the tour represents an image of Africa and that the white saviour can enact their role. Mary, who helps to distribute Adventure Company's funding, also plays a role in the production of the movie. Her role, however, does not fit as neatly into the role of 'producer' – as she worked behind the scenes to help determine how the Adventure Company tour would look, but also how the funding was implemented. Perhaps, in this metaphor she could be considered to play the role of an acting coach – as she worked behind the scenes to ensure all roles were played correctly. Greg's role in the movie metaphor has previously been explored in this chapter and in Chapter 5.

The interview with Mary represents, in part, a complexity and an anomaly in the research data. Mary's interview is reflective of the multitude tensions present in Zimbabwe, tensions from the land reforms of the early 2000s, the Independence wars, and colonisation, to name a few. There are well-established identity politics at work in Zimbabwe, as white Zimbabweans navigate their positioning in a post-colonial Zimbabwe (Tagwirei, 2016; Tagwirei & de Kock, 2015). Musanga (2016) refers to these tensions as 'evocative of the psychological and physical insecurity pervading the white community' (p. 401). Alexander (2004) describes these identity politics as '[white Zimbabweans] are, in essence, "Orphans of the Empire"; while they live in and love Zimbabwe, they do not feel that they are considered native' (p. 196). Other scholars have discussed the ways in which paternalism is imbued in the talk of farmers and the issues for white Zimbabweans regarding belonging and identity (Kalaora, 2011; Pilossof, 2009; Suzuki, 2017, 2018). Race, and racism, is a prevalent issue in Zimbabwe and it would be remiss to ignore (Wylie, 2012).

Part of the troubling aspect of Mary's interview was in her discussion of Matopo School. Mary's interview occurred late in the fieldwork, and I had

formed a strong connection with the school. In reading the interview tran-
script, it is hard to match my own understanding of Matopo School with
Mary's understanding of Matopo School. Mary's understanding is, in part,
due to her positioning. Mary's positioning is reflective of the identity politics
at work and highlights the complexities of researching in a postcolonial space.
In what follows, I try to unravel the anomaly that Mary represents in the
research data, but also reflect on the white community that surrounds
the school.

In her interview, Mary articulated doubts about working with the local,
black community that surrounds Matopo School:

> I was not keen on doing community projects there at all. I just did not get
> a feeling. A good feeling about it. I have been forced into it by Thompson
> and [Adventure Company personnel] together. But again, the way I run
> programs, is they have to give their all. And if they do not, the program
> will not flourish. It will tick along, but then I will move out and they have
> got to carry on.

When asked why she did not have a good feeling about the community, she said,

> I do not know. It just was an underlying sort of underlying feeling. That
> they were.... [**pause**] they were going to be hard work, yeah.

Mary's representation of the community is reflective of the moral economy
discourse – which promotes hard work as a reason for success. According to
Wale and Foster (2007) the moral economy discourse:

> Supports the notion that the wealthy get jobs based on the fact that they
> are hard workers, and hard workers are morally good people; therefore,
> the wealthy are morally good people deserving of their wealth. In oppo-
> sition, the poor are poor because they are lazy sinners.
>
> (p. 52)

The moral economy discourse can be seen in Mary's insistence that the community need to be hard workers for her to continue to fund them. Further, the
articulation of needing to conform to white ways of knowing and doing is
endemic in development (Baptista, 2017; Escobar, 1995/2012; Goudge, 2003;
Quijano, 2010). In some ways, Mary is constrained by the very field in which
she works – as the dominant discourses about how Development *should be*
frame her relationships and interactions. Mary later explained she does not
work with another school:

> With [School Name], I am pulling a little bit out, because I find she is too
> demanding and needy. And the minute you do that, I cannot handle it.

By being 'too demanding and needy' the other school no longer receives funds from the charitable organisation Mary works for. In this case, it could be argued the other school is constituted as 'doing' Development wrong, as they are no longer passive recipients of the aid. Silk (2004) identifies that in dispersal of aid there is a construction of those in the Minority World as 'active and generous' and those in the Majority World as 'cared for, passive and grateful' (p. 230). As I explored earlier in this chapter, the practice of gift-giving governs those constituted as 'donor' and those who are constituted as 'recipient'. Mary, then, is placed in a difficult position in her role as mediator between donor and recipient.

In Mary's role, as intermediary between donor and community and in as the distributor of funds, she has sustained engagement with some communities. She differentiates these communities from the 'needy' and 'demanding':

> [this does not happen in] our communities where we work a lot, who know me really well. They know that I do not come from a wealthy family, they know that I had to pay for my own education.

By making this assertion, 'I had to pay for my own education', Mary appeals to the hard work ethos of the moral economy discourse. In a study of a white farming community, Suzuki (2017, 2018) identified that articulations of working hard to buy their land fairly was a common discursive practice by white farmers and Suzuki concludes it is an attempt to distance themselves from colonial histories of inequality. Despite not coming from a wealthy family, Mary was educated during the colonial period; thus Mary was privileged with easier entry to higher levels of education than her black African counterparts (as briefly explored in Chapter 2). In describing her own background as *not* wealthy, Mary is discursively distancing herself from the laws and policies that were in place, which have contributed in some way to her current life position, whether she chooses to acknowledge this or not. The invisibility of whiteness allows the privilege of being white to usually be invisible to those who are white (Goudge, 2003; Suzuki, 2017, 2018; Tagwirei & de Kock, 2015; Villet, 2012).

In the extract from Mary's interview, she acknowledges the significance of her white skin and the connotations this has with wealth. By working with communities and allowing them to see that she does not come from wealth, she is trying to remove the presence of whiteness as a marker of wealth. Mary in some way recognises whiteness as a governing ideal in Zimbabwe. Previous studies have established that whiteness is both an invisible and visible marker in Zimbabwe (Suzuki, 2017, 2018). In some ways, however, the privilege of white skin is invisible to white Zimbabweans, such as through the wealth accumulation that results from colonisation (Pilossof, 2009; Suzuki, 2017, 2018).

For Thompson, the black Zimbabwean school founder, the privileging of whiteness had been explicit throughout his life:

It is 10 times harder, in any respect. Even at school, if you are [**pause**] a black little kid you have got to work twenty times harder. If you are working, here in Africa with a white counterpart you have got to work tenfold. Because there is the assumption that this white guy is already good. You have to prove you are good.

(Thompson)

The inherent privilege of whiteness creates complexities in the implementation of tourism and development work. When asked whether it would be easier for him if he was white, Thompson replied,

Yeah. Because if it was a white guy walking into an office, you know he is offered a chair, it is already 'how can I help you, *SIR*.' Whereas I get, 'what do you want?' It is actually tougher. Because you do not get respected. And they think, 'who are you? To think you can do this?' Nobody believed me when I said, 'I want to build a school'. The whole community said, 'just forget about this crazy guy from Masvingo, he does not know what life is all about'.

He further described the problems he had encountered:

People disregard each other, out of [**pause**]… it is just the way things were taught, and the way life was believed to be and it stayed that way for so long. Go to the shop that we buy electrical goods from, go there and see how many people will come to you and say, 'can I help you madam?' There is a local guy who probably is going to buy more, nobody cares about him because it is just the mentality that, I have to go and help that person because they are white. And the local guy will see himself out and they say, 'he probably isn't buying anything anyway'. I go there, buy things that are valued to about $2000 a day, I still have people that will not say good morning to me. In the same shop. They know I am paying their salary, but they will not say good morning, because I am not important.

For Thompson there are strong colonial relics in his interactions with both the local black community surrounding the school and the white community who live in a wealthier suburb nearby. As he says, *it is just the way things were taught*, and *the way life was believed to be*. In the postcolonial landscape of Zimbabwe 'assumptions about racial difference constitute an interpretive grid through which people make sense of everyday experience' (Suzuki, 2017, p. 75). Thompson's interview makes it explicit that whiteness, and race, is still a formidable structuring force in Zimbabwe. In his interview, Thompson identified the ways in which modern racial formations are part of the complexities in the implementation of tourism in the school.

These racial formations played out in the way that Mary and Thompson spoke about each other. In her interview, Mary discussed Thompson with me, asking me if I knew why Alison and Thompson had begun working together:

| | |
|---|---|
| *Mary:* | Be interesting to know how, but you can ask her that! And *why* Thompson. |
| *Interviewer:* | I think he can be quite charismatic - |
| *Mary:* | [interrupts] Oh, no! He is *lovely*. That is what I was going to say. But why? I mean charismatic... but great that he has got it. And I love the fact that he has put it into a school. But what made them... well maybe he is the only person they have met in Africa? |

To contextualise the comments in the extract, Thompson met Alison and her husband when he worked as tour guide for them, and then on a few more occasions. Together they have an established professional relationship, through which they have created and developed a tour company (Shumba Safaris), Matopo School, and a medical clinic. Stereotypes and dominant discourses position black Africans, particularly men, as out to trick or fool tourists or potential donors (Martins, 2011; Salazar, 2009), as explored in Chapter 5. Thompson has had significant success as a private tour guide and the comment of *why Thompson* could be interpreted as dismissive of this success. In this sense, it is striking that the discourses about the dangers of African men have such authority that a white Zimbabwean woman also plays into them. On the other hand, these discourses also constitute both Alison and her husband as foolish or having been tricked by Thompson in some way as the power relations constitute white tourists in a particular way. This is visible when Mary concludes *maybe he is the only person they have met in Africa*. It is important to note two things here: Alison and Thompson initially started a tourism company, something which Thompson was qualified to do. Second, both initiatives form some sort of threat to the work Mary conducts as they are in direct competition with the tours that she helps to organise through Adventure Company, and the work of the charitable organisation she is employed by.

The assertion 'he is *lovely*' is an example of 'dilemmas of interpretation' (Durrheim et al., 2011, p. 35). While previous studies of race in Zimbabwe have identified explicit examples of racism (Suzuki, 2017; Wylie, 2012), 'dilemmas of interpretation' contribute a new understanding to the ways in which whiteness plays out. Durrheim et al. (2011) detail this,

Black people are often confronted with profound doubt about whether they are the victims of racism because there is often uncertainty about whether an incident is racist or not... This uncertainty about racism is no mere coincidence. Racial expressions are often specifically designed to allow people to deny that they are being racist.

(p. 36)

Mary's question of *why Thompson* followed by the assertion, *he is lovely*, could contribute to a dilemma of interpretation in this case. One interpretation may be that, indeed, Mary does think that Thompson is suitably qualified. On the other hand, Durrheim et al. (2011) suggest that these forms of denials are part of the dilemmas of interpretation that work to protect comments based on race. Mary's interview included a few instances of dilemmas of interpretation, where the racialisation was not explicit and in fact *might not* be there at all. This sense of ambiguity and contestable nature of the interpretation is reflective of modern racial formations; which Durrheim et al. (2011) argue allow racism to become insidious within South Africa today, as it ensures that any statement can be passed off as opinion, and not about race.

It was clear in Thompson's interview that, for him, there were no dilemmas of interpretation – in any situation – involving race in Zimbabwe:

> If my project was done by some of the local white guys, locally *all* the whole community would be there. They would go there and help you know 'we have come to support you; can we help you?' But because it is Thompson, the locals will not help me.… [so I have learnt] you just do your work, do not wait for praise or do not wait for comments from people. You will not get anywhere.

Thompson does not feel supported by the local white community. In my experience, when meeting other white community members, many did not know that Thompson's school existed. Instead, they often assumed I meant a school in a nearby area, financed by a local white woman. Thompson was quite frustrated by the situation:

> Can you imagine if we are getting even a quarter of that help? From the locals, saying, 'Hey did you know that there is a school just outside of town, started by one of the guides?' No. Umm [they say instead], 'that one. It is just a once off case in Africa'. So, no, people do not care.

Thompson, expressed that there was even a level of mistrust from the local community:

> No, no, they just think you have got a crazy ideas and that you are probably using most of the money for yourself, so, what is the point in helping?
> (Thompson)

In the interviews with Mary and Greg, they did not express overtly negative opinions or racist perspectives. But it is clear from Thompson's interview that he feels there is a level of injustice occurring, and he attributes it to race. Of interest here, is Thompson's assertion *just a once off case in Africa*. Discursively, it has echoes of Mary's term *he is the only one they have met in Africa*. For Thompson, race is a structuring force which limits his ability to have success or engage with the local white community. Clausen (2019) identifies that NGOs are 'often accused of preserving the status quo by setting up systems of patronage that undermine and depoliticise social movements and other grassroots organisations' (p. 80). In this case, the local white community continues to preserve the status quo of white-dominated support and aid. To put it in Foucauldian terms, the dominant discourses frame Thompson's work as unusual and a 'once-off', but they also constitute Thompson as potentially looking to steal money or defraud the local community (Martins, 2011; Salazar, 2009). In itself, this is further complicated in that Adventure Company and the charitable organisation Mary works for is multinational – it is an international NGO undermining the grassroots organisation developed by Alison and Thompson.

As I have demonstrated in this section, whiteness and race were ever-present in the interactions between the stakeholders – the tourism does not occur in a vacuum and the local white community have a role to play. I found this section of the book the hardest to write, and to write well. It has undergone multiple iterations and formats, as I grapple with the question 'how much responsibility do I have towards the words of my research participants?' that is, how much do I owe to Mary and my representation of her – given she participated in the research as a consenting adult? How much of what she said comes down to 'dilemmas of interpretation'? I hope that I have represented Mary in a fair manner. Her interview was troubling at times but is deeply reflective of her own positioning within the modern race formations in Zimbabwe. She had a level of care for the children at Matopo School and she was trying her best to provide an education for them – in the manner that she knew how.

**Summary**

First, in this chapter I developed the idea of the 'star' in this metaphor, the white saviour. The white saviour is a role which is unquestioned and supported by colonial imageries of the 'dark continent'. In this section I used examples of the ways in which the tourists constructed their own actions within the space, to show how preconceived notions of philanthropy are reinforced by discourses of the white saviour. The discursive constructions which are perpetuated by the white saviour discourses are harmful, as they continue to structure world systems and reinforce the privileging of Minority World conceptions of modernity. Following the argument of Bell (2013), there is a relationship between the white saviour and the Other – one cannot exist without the other. I explored my relationship and interactions with the tourists, demonstrating how they placed me within their known discursive framework: that of the white saviour.

Second, I reflected on one disagreement between the three main stakeholders, Mary, Thompson, and Alison. Each of these people work from within their own positioning as they navigate the challenges of implementing funding in a school. This tension highlights the tensions that are ever-present in projects funded by external funders.

Last, in this chapter I have reflected on the role the local white community has in relation to the school. As identified by Thompson, this role is very little, but he believes it could be more. There were 'dilemmas of interpretation' which played out in the interviews with the participants. This is reflective of the modern racial formations in Zimbabwe and these dilemmas highlight the way in which black Zimbabweans are still subjugated by white communities.

## Note

1 I recognise that people of colour are always already marked by their skin. For me, however, this fuelled an increasing frustration towards the tourists. This is to explain the frustrated/angry tone with which some of the field notes are tinged.

## References

Achebe, C. (1998). An image of Africa: Racism in Conrad's Heart of darkness. In A. M. Roberts (Ed.), *Joseph Conrad*. Addison Wesley Longman Limited.

Alexander, K. (2004). Orphans of the empire: An analysis of elements of white identity and ideology construction in Zimbabwe. In B. Raftopoulos & S. Tyrone (Eds.), *Zimbabwe: Injustice and Political Reconciliation* (pp. 193–212). Weaver Press.

Ansell, N. (2008). Third world gap year projects: Youth transitions and the mediation of risk. *Environment and Planning D: Society and Space*, 26(2), 218–240. https://doi.org/10.1068/d83j

Arnett, J. (2016). Taking pictures: The economy of affect and postcolonial performativity in NoViolet Bulawayo's We Need New Names. *Ariel: A Review of International English Literature*, 47(3), 149–173.

Bandyopadhyay, R. (2019). Volunteer tourism and "The White Man's Burden": Globalization of suffering, white savior complex, religion and modernity. *Journal of Sustainable Tourism*, 27(3), 327–343. https://doi.org/10.1080/09669582.2019.1578361

Bandyopadhyay, R., & Patil, V. (2017). 'The white woman's burden' – the radicalized gendered politics of volunteer tourism. *Tourism Geographies*, 19(4), 644–657. https://doi.org/10.1080/14616688/2017.1298150

Baptista, J. A. (2017). *The good holiday: Development, tourism and the politics of benevolence in Mozambique*. Berghahn Books.

Bargeman, B., Richards, G., & Govers, E. (2018). Volunteer tourism impacts in Ghana: A practice approach. *Current Issues in Tourism*, 21(13), 1486–1501. https://doi.org/10.1080/13683500.2015.1137277

Bauer, I. (2017). More harm than good? The questionable ethics of medical volunteering and international student placements. *Trop Dis Travel Med Vaccines*, 3(5), 5. https://doi.org/10.1186/s40794-017-0048-y

Bell, K. M. (2013). Raising Africa? Celebrity and the rhetoric of the white saviour. *PORTAL Journal of Multidisciplinary International Studies*, 10(1). https://doi.org/10.5130/portal.v10i1.3185

Bex, S., & Craps, S. (2016). Humanitarianism, testimony, and the white savior industrial complex: What is what versus Kony 2012. *Cultural Critique*, 92, 32–56.

Bhambra, G. K. (2013). The possibilities of, and for, global sociology: A postcolonial perspective. In J. Go (Ed.), *Postcolonial sociology, Political power and social theory* (Vol. 24, pp. 295–314). Emerald Group Publishing Limited.

Briggs, J., & Sharp, J. (2006). Indigenous knowledges and development: A postcolonial caution. *Third World Quarterly*, 25(4), 661–676. https://doi.org/10.1080/01436590410001678915

Cammarota, J. (2011). Blindsided by the Avatar: White saviors and allies out of Hollywood and in education. *Pedagogy and Cultural Studies*, *33*(3), 242–259. https://doi.org/10.1080/10714413.2011.585287

Chouliaraki, L. (2010). Post-humanitarianism Humanitarian communication beyond a politics of pity. *International Journal of Cultural Studies*, *13*(2), 107–126. https://doi.org/10.1177/1367877909356720

Christian, M. (2019). A global critical race and racism framework: Racial entanglements and deep malleable whiteness. *Sociology of Race and Ethnicity*, *5*(2), 169–185.

Clausen, H. B. (2019). NGOs, tourism and development. In R. Sharpley & D. Harrison (Eds.), *A research agenda for tourism and development* (pp. 71–87). Edward Elgar Publishing Limited.

Cohen, S. A., & Cohen, E. (2017). New directions in the sociology of tourism. *Current Issues in Tourism*, *22*(2), 153–172. https://doi.org/10.1080/13683500.2017.1347151

Cravatte, C., & Chabloz, N. (2008). Enchantment and solidarity. *Tourist Studies*, *8*(2), 231–247. https://doi.org/10.1177/1468797608099250

Crossley, É. (2012). Poor but Happy: Volunteer Tourists' Encounters with Poverty. *Tourism Geographies*, *14*(2), 235–253. https://doi.org/10.1080/14616688.2011.611165

Durrheim, K., Mtose, X., & Brown, L. (2011). *Race trouble: Race, identity and inequality in post-apartheid South Africa*. Lexington Books.

Escobar, A. (1984). Discourse and power in development: Michel Foucault and the relevance of his work to the third world. *Alternatives*, *10*(3), 377–400.

Escobar, A. (1995/2012). *Encountering development - The making and unmaking of the Third World*. Princeton University Press. (Paperback reissue 2012).

Faria, C., & Mollett, S. (2014). Critical feminist reflexivity and the politics of whiteness in the 'field'. *Gender, Place & Culture*, *23*(1), 79–93. https://doi.org/10.1080/0966369x.2014.958065

Foucault, M. (1975/1991). *Surveiller et punir: Naissance de la prison [Discipline and Punish: The Birth of the Prison]* (A. Sheridan, Trans.). Penguin Books Australia Ltd.

Foucault, M. (1976/2008). *La Volonté de-savoir [The History of sexuality: Volume 1]* (R. Hurley, Trans.). Penguin Group (Australia).

Foucault, M. (1982/2002). The subject and power (Robert Hurley and others, Trans.). In J. D. Faubion (Ed.), *Dits et ecrits 1954–1984 [Essential works of Foucault 1954–1984]* (Vol. 3: Power, pp. 326–348). Penguin Books Ltd.

Frazer, R., & Waitt, G. (2016). Pain, politics and volunteering in tourism studies. *Annals of Tourism Research*, *57*, 176–189. https://doi.org/10.1016/j.annals.2016.01.001

Garland, E. (2012). How should anthropologists be thinking about volunteer tourism? *Practicing Anthropology*, *34*(3), 5–9.

Godfrey, J., Wearing, S., & Schulenkorf, N. (2015). *Neo-colonialism and the volunteer tourist gaze: Commerical volunteer tourism in Cusco, Peru* CAUTHE 2015: CAUTHE 2015: Rising Tides and Sea Changes: Adaptation and Innovation in Tourism and Hospitality, Gold Coast, QLD: School of Business and Tourism, Southern Cross University. https://search.informit.com.au/documentSummary;dn=208600243651280;res=IELBUS

Goudge, P. (2003). *The Whiteness of Power: Racism in Third World Development and Aid*. Lawrence & Wishart Ltd.

Harper, M., & Smithers, K. (2023). Fieldwork from A–Z? Exploring shifting identities in doctoral research in Australia and Zimbabwe. *The Australian Educational Researcher*. https://doi.org/10.1007/s13384-023-00649-1

Harrison, D. (2015). Development theory and tourism in developing countries: What has theory ever done for us? *International Journal of Asia Pacific Studies*, *11*(S1), 53–82.

Hernandez, K. (2018). Can the subaltern be seen? Photographic colonialism in service learning. *Qualitative Research Journal*, *18*(2), 190–197. https://doi.org/10.1108/QRJ-D-17-00051

Hughey, M. W. (2010). The white savior film and reviewers' reception. *Symbolic Interaction, 33*(3), 475–496. https://doi.org/10.1525/si.2010.33.3.475

Huxley, L. (2004). Western backpackers and the global experience: An exploration of young people's interaction with local cultures. *Tourism, Culture & Communication, 5,* 37–44.

Jackson, L. (2013). They don't not want babies: Globalizing philosophy of education and the social imaginary of international development. In C. Mayo (Ed.), *Philosophy of Education* (pp. 353–361). Philosophy of Education Society.

Jakubiak, C. (2016). Ambiguous aims: English-language voluntourism as development. *Journal of Language, Identity & Education, 15*(4), 245–258. https://doi.org/10.1080/1 5348458.2016.1195270

Judge, R. C. (2016). Negotiating blackness: Young British volunteers' embodied performaces of race as they travel from Hackney to Zimbabwe. *YOUNG Editorial Group, 24*(3), 238–254. https://doi.org/10.1177/110330881562635

Judge, R. C. (2017). Class and global citizenship: Perspectives from non-elite young people's participation in volunteer tourism. *Tourism Recreation Research, 42*(2), 164– 175. https://doi.org/10.1080/02508281.2017.1303894

Kalaora, L. (2011). Madness, corruption and exile: On Zimbabwe's remaining white commerical farmers. *Journal of Southern African Studies, 37*(4), 747–762. https://doi. org/10.1080/03057070.2011.609341

Kapoor, I. (2004). Hyper-self-reflexive development? Spivak on representing the third world 'Other'. *Third World Quarterly, 25*(4), 627–647. https://doi.org/10.1080/014365 90410001678898

Knaus, C. B. (2019). "If everyone would just act white": Education as a global investment in whiteness. In J. S. Brooks & G. Theoharis (Eds.), *Whiteucation: Privilege, power and prejudice in school and society* (pp. 1–21). Routledge.

Kobayashi, A., & Peake, L. (2000). Racism out of place: Thoughts on whiteness and an antiracist geography in the new millennium. *Annals of the Association of American Geographers, 90*(2), 392–403. https://doi.org/10.1111/0004-5608.00202

Kothari, U. (2006). An agenda for thinking about 'race' in development. *Progress in Development Studies, 6*(1), 9–23. https://doi.org/10.1191/14649963406ps124oa

Lorenzini, D. (2015). What is a "Regime of Truth"? *Le foucaldien, 1*(1). https://doi. org/10.16995/lefou.2

Lough, B. J. (2015). Confronting the white elephant: International volunteering and racial (dis)advantage. *Progress in Development Studies, 15*(3), 207–220. https://doi. org/10.1177/1464993415578983

Lozanski, K. (2010). Independent travel: Colonialism, liberalism and the self. *Critical Sociology, 37*(4), 465–482. https://doi.org/10.1177/0896920510379443

Martins, C. (2011). The dangers of the single story: Child-soldiers in literary fiction and film. *Childhood, 18*(4), 434–446. https://doi.org/10.1177/0907568211400102

Matiza, T., & Oni, O. A. (2014). Managing the tourist destination image: The case of Africa. *Tourism Review, 62*(4), 396–406.

Monshausen, A., Plüss, C., & Maurer, M. (2015). *From volunteering to voluntourism: Challenges for responsible development of a growing travel trend.* B. f. t. World. https:// www.tourism-watch.de/files/profil18_voluntourism_final_en.pdf

Mostafanezhad, M. (2013a). The geography of compassion in volunteer tourism. *Tourism Geographies, 15*(2), 318–337. https://doi.org/10.1080/14616688.2012.675579

Mostafanezhad, M. (2013b). 'Getting in touch with your inner Angelina': Celebrity humanitarianism and the cultural politics of gendered generosity in volunteer tourism. *Third World Quarterly, 34*(3), 485–499. https://doi.org/10.1080/01436597.2013.7 85343

Mostafanezhad, M. (2013c). The politics of aesthetics in volunteer tourism. *Annals of Tourism Research, 43,* 150–169.

Mowforth, M., & Munt, I. (2009). *Tourism and sustainability: Development globalisation and new tourism in the Third World* (3rd ed.). Routledge.

Musanga, T. (2016). White Zimbabwean farmers' unstable mobilities identity and history in Douglas Rogers' the last resort: A memoir of Zimbabwe. *National Identities, 18*(4), 397408. https://doi.org/10.1080/14608944.2015.109182

Pilossof, R. (2009). The unbearable whiteness of being: Land, race and belonging in the memoirs of white Zimbabweans. *South African Historical Journal, 61*(3), 621–638.

Power, R., Langhaug, L. F., Nyamurera, T., Wilson, D., Bassett, M. T., & Cowan, F. M. (2004). Developing complex interventions for rigorous evaluation - a case study from rural Zimbabwe. *Health Education Research, 19*(5), 570–575. https://doi.org/10.1093/her/cyg073

Quijano, A. (2010). Coloniality and modernity/rationality. In W. D. Mignolo & A. Escobar (Eds.), *Globalization and the decolonial option* (pp. 22–32). Routledge.

Salazar, N. B. (2009). Image or imagined? Cultural representations and the "tourismification" of peoples and places. *Cahiers détudes Africaines, 49*(193–194), 49–71. https://doi.org/10.4000/etudesafricaines.18628

Shelab, M. (2011). *Tourism-led development in South Africa: A case study of the Makuleke partnership with Wilderness Safaris.* University of Witwatersrand.

Sheridan, V., Landy, D., & Stout, V. (2019). The return of the 'Black Babies': How development education affects schoolchildren's attitudes to the majority world - the Trócaire 2012 Lenten campaign. *Race, Ethnicity and Education, 22*(6), 858–874. https://doi.org/10.1080/1361332.2017.1417829

Silk, J. (2004). Caring at a distance: Gift theory, aid chains and social movements. *Social & Cultural Geography, 5*(2), 229–251. https://doi.org/10.1080/14649360410001690231

Snee, H. (2013). Framing the Other: Cosmopolitanism and the representation of difference in gap year narratives. *The British Journal of Sociology, 64*(1), 142–162. https://doi.org/10.1111/1468-4446.12010

Steyn, M. E. (2001). *"Whiteness just isn't what it used to be": White identity in a changing South Africa.* State University of New York Press.

Suzuki, Y. (2017). *The nature of whiteness: Race, animals, and nation in Zimbabwe.* University of Washington Press.

Suzuki, Y. (2018). The good farmer: Morality, expertise, and articulations of whiteness in Zimbabwe. *Anthropological Forum, 28*(1), 74–88. https://doi.org/10.1080/00664677.2018.1429252

Tagwirei, C. (2016). The nucleation of white Zimbabwean writing. *Journal of Literary Studies, 32*(3), 5–20. https://doi.org/10.1080/02564718.2016.1235377

Tagwirei, C., & de Kock, L. (2015). Whitelier than white? Inversions of the racial gaze in white Zimbabwean writing. *African Identities, 13*(3), 184–198. https://doi.org/10.1080/14725843.2015.1023256

Toomey, N. (2017). Humanitarians of Tinder: Constructing whiteness and consuming the Other. *Critical Ethnic Studies, 3*(2), 151–172.

Villet, C. (2012). The invisibility of richness: A critique of Vice's 'strange place'. *South African Journal of Philosophy, 31*(4), 702–716. https://doi.org/10.1080/02580136.2012.10751803

Wale, K., & Foster, D. (2007). Investing in discourses of poverty and development: How white wealthy South Africans mobilise meaning to maintain privilege. *South African Review of Sociology, 38*(1), 45–69. https://doi.org/10.1080/21528586.2007.10419166

Wearing, S., Mostafanezhad, M., Nha, N., Nguyen, T. H. T., & McDonald, M. (2018). 'Poor children on Tinder' and their barbie saviors: Towards a feminist political economy of volunteer tourism. *Leisure Studies, 37*(5), 500–514. https://doi.org/10.1080/02614367.2018.1504979

Wylie, D. (2012). Not quite a refutation: A response to David McDermott Hughes's *Whiteness in Zimbabwe*. *Safundi*, *13*(1–2), 181–194. https://doi.org/10.1080/17533171.2011.644453

Yancy, G. (2004). Introduction: Fragments of a social ontology of whiteness. In G. Yancy (Ed.), *What white looks like: African-American philosophers on the whiteness question* (pp. 1–24). Routledge.

# 8    Conclusion

In early 2020, COVID-19 caused mass lockdowns and a significant number of deaths around the world. At the time, a number of tourism scholars reckoned with the significance of such a global event for the tourism industry and for the communities involved. It was suggested at the time that the global pandemic could act as a reset button, particularly in creating the conditions for a reconsideration of the effects of overtourism and the significant ecological impacts of mass tourism. Unfortunately, this did not happen. School tours continue, volunteer tourism continues, orphanage tourism continues. These tourisms have continued for years despite calls from critical tourism scholars and news media reporting about their mostly detrimental effects on the communities who are visited.

Somewhat paradoxically, although these activities still exist the scope and size of school tour tourism remains unclear. In Zimbabwe, it often occurs in an ad-hoc manner with individual tourism providers organising tours and philanthropic donations. It is reported that 95% of tourism providers in Matabeleland North perform some form of philanthropy, such as school fee donations or donations of larger infrastructure. As outlined in Chapter 3, Matopo School has formal arrangements with two multinational tourism companies – Education Tours and Adventure Company. The partnerships at Matopo School highlight that this tourism is not just occurring in ad-hoc arrangements, but rather there are large tourism companies enacting this type of tourism. In considering the design of Matopo School as a grassroots, community-driven response to the needs of the community, it is interesting to note that what began at a small scale now has large companies involved. The arrangements at Matopo School add an additional dimension to what we know about this type of tourism, in highlighting that the arrangements of funding may be a combination of both grassroots philanthropy and larger corporate social responsibility initiatives drive by large charitable organisations.

School tours could fit within many types of tourism. What I have hoped to show throughout this book is that school tours straddle an uneasy line between small-scale sustainable tourism venture and mass tourism. They are at once an on-the-ground mechanism that can begin to alleviate the economic struggles of schools in Zimbabwe, but are also a pre-arranged venture that is marketed

DOI: 10.4324/9781003407980-8

on-mass to Americans over the age of 50. In some respects, school tours might be considered the older American's answer to volunteer tourism or the Peace Corps, in that they enable tourists to enact their white saviour visions whilst participating in a pre-organised activity.

There is no doubt that this form of tourism, and others like it, mirror and reflect the longer traditions and flows of Development and development discourses. In Chapter 4, I outlined the history of development and the tourisms that seem to proliferate and continue in its likeliness. These tourisms are deeply problematic in the way that they encourage the image of the white saviour. It is not just tourisms that encourage images of the white saviour, it is also celebrity humanitarianism and popular charitable ventures. These images are ubiquitous and enduring.

As outlined in Chapter 4, there are multiple and competing images of Majority World children. These images are prolific in their reach and magnitude. The type of tourism I have written about across this book is mostly based on the image of the poor-but-happy child. This imagery obscures the lived effects of poverty for host communities, as tourists are able to excuse the conditions under which people live because they appear to be 'happy'. These images also work to appease feelings of guilt or uncertainty by the tourists. The poor-but-happy child extends into not just tourism, but on social media and through other charitable initiatives.

As I outlined in Chapter 2, the economic history of Zimbabwe somewhat explains the educational present, in which there is significant under-resourcing of schools. Zimbabwe is still in a period of hyperinflation and is still responding to the aftereffects of the 2007/2008 economic collapse, which included a 'brain drain' of qualified teachers. Government schools, particularly those in rural areas, lack basic facilities. As a mechanism to resolve the shortages of classrooms and desks, satellite schools were introduced to enable local communities to run and finance their own schools. Matopo School is one such school, through which charitable donations fund most of the school infrastructure. In the case of Matopo School, this satellite arrangement has worked well. They have electricity, flush toilets, a lunch programme, and access to a computer lab. It is not hyperbolic to say this is a rarity for rural schools in Matabeleland North, with the statistics showing that in 2018, only 18% of primary schools had an electricity source and only 10% had internet connectivity (Ministry of Primary and Secondary Education, 2019).

In this book I set out to examine the taken-for-granted image of Africa that is proliferated in the heterotopia of Matopo School. To do so, I used the metaphor of a movie, with set design, actors, producers, and directors. Across the last few chapters I have outlined the conflicting demands for the different roles at the school. These conflicting demands shape the implementation of the tourism and have the potential to offer insight into how school tours could look.

In the metaphor, both the school founders and tourism personnel act as 'producers' of the tourism in the school. In their roles, they work to oversee a broader vision of the school tour and manage the 'set' design. Ultimately, the school founders are responsible for the success and management of the school, including the running of the school tour. As part of this role, they are responsible for the creative vision and direction of the school tour. The tourism personnel, while also 'producers', work from a different perspective. For example, Greg, a tour guide, outlined in Chapter 5:

> After the trip we get a report back and the guests are always asked what part of the trip is the best part. What I try and make sure I get is, 'I cannot pinpoint anything because every day had something to offer'.

The extract from Greg highlights a key argument of this book that the teachers and tourism personnel work from different and conflicting perspectives: an educational perspective and a business perspective. One of the main aims for Thompson and Alison, as school founders, is to increase the education standard at Matopo School – although they understand the need to provide a tourist product that is marketable and fits a particular image of Africa. However, for the tourism personnel, such as Greg, the aim is to get a good report at the end of the trip – with the school tour being just one element of the overall itinerary.

Chapter 5 argued that as producers, the tourism personnel and school founders actively work to create a heterotopia built from the image of Africa. This heterotopia includes specific imagery associated with the image of Africa which plays out in marketplaces, landscapes, and safaris. In Chapter 5 the work of Greg was examined in depth, with particular reference to a marketplace visit he would conduct as part of his itinerary. In Chapter 6, the image of the tourists as needing interaction with children was explored. Discourses of love and care constitute the school tour in such a manner that having a tour *without* children seems almost impossible for the tourism personnel. Further, as their work depends on a positive report at the end of the trip, rather than the educational outcomes of children, it could be argued the tourism personnel have very little need to consider the educational impacts at all.

The work of Mary, one of the tourism personnel, was heavily constrained by the dominant imagery of children needing love, care, and protection. Her role was unique, as she also ran educational programmes in schools across Matabeleland North for a charitable organisation associated with Luxury Tours. However, when I spoke to Mary, her role at Matopo School was only in the implementation of funding for Adventure Company's charitable organisation. As a qualified teacher, in her interview she discussed at length the educational impacts of school tours. She had devised a solution, which would allow tourists to interact with children while engaging in an educational activity, although it had not been fully implemented at the time of the interview. Her role at Matopo School, at the time of the interview, was framed through her experience as an educator but constrained by the wishes of the donors (as explored in Chapter 7). For Mary, someone with an education background, a school tour without interaction with children also seemed almost impossible, due to the needs and wishes of donors.

The school founders, who are also tourism personnel, have a multifaceted role as producers of the movie. As demonstrated in Chapter 5, they are aware of the image they are required to cultivate to encourage tourists to visit. As explored further in Chapter 6, they are also unsure if the school tour could run without interaction with children. As shown in Chapter 7, their relationship with the other tourism companies who visit the school is sometimes fraught due to competing wishes and demands. In terms of their work in the school, the school founders demonstrated a consideration of the ways the school was affected by the tourism; however, they were constrained by their dual role as business owners and school founders. As business owners, Thompson and Alison were responsible for maintaining a successful safari business – a company whose clients who also visited the school and provided substantial donations. The funding arrangement for the school relies heavily on private contributions by their safari company and its guests; this means the school could not continue in the same manner without the successful safari business. As school founders, Thompson and Alison were responsible for running and maintaining a school in a way that would provide an optimal educational environment for the children in their care. Thompson and Alison navigated this tension in many ways, acknowledging the need to provide guests with what they want, but also identifying that as children get older the school tour may look different.

The teachers are the 'directors' of the film set, in which they perform a more direct role of portraying the image of Africa required for the tour. I argue the functioning of the school is disrupted by tourism; however, the teachers justify the interruptions as the tourism provides positive benefits to the school. During the school tour, classes are interrupted, and teachers must leave their classrooms. The teachers navigate conflicting roles as they manage these disruptions, which requires them to move between the role of 'good' teacher and the role of a 'good' tour guide who must produce an ideal 'image of Africa' on the school tour.

Interestingly, the teachers reported how they felt the school tour could change; however, this was constrained by the dominant expectation of conducting the tour 'well'. Conducting the tour 'well' might mean conforming to the expectations of what it means to be an 'African' school. The teachers and tourism personnel were concerned that if the school tour did not involve children, tourists would not want to engage with the tour (as detailed in Chapters 5 and 6). This concern is representative of the tension the teachers felt between conducting the tour 'well' and the need to educate children in the school. Further, in their role of 'director' they manipulate the role of the children in the school tour, to enhance the feelings of love (Chapter 6) for children and feelings of trust in the school by the tourists (Chapter 5). In many ways, the teachers perceived their role as someone who had to enhance the enjoyment of the tour, in order to produce further funding for the school.

At the very core of all these activities and tourisms is the notion of 'race trouble'. In Chapter 5 I identified how the teachers in this study discussed a feeling of mistrust from the tourists. The perceived feeling of mistrust demonstrates the ways in which the teachers 'gaze' back to the tourists as they negotiate and shape the tour around these expectations. Further, the perceived mistrust is reflective of a colonial discourse which constitutes all people who live in African countries as scheming to defraud or thieve from Minority World tourists (Aitchson, 2001). One example of the mistrust was demonstrated in an example given by the teachers, in which they were asked to provide evidence that donations had been spent correctly. The mistrust was further demonstrated by the teachers who altered parts of the tour to 'show' the tourists they are honest, such as the children participating in beading, but only during the school tours.

As a remedy to the perceived mistrust, the teachers encourage a positioning of the children in the school, that of the poor-but-happy child. Minority World images of children are imbued with colonial discourses of care, in which they are positioned as poor-but-happy or fly-blown and poverty stricken, a very familiar image. The school adds to the image of the poor-but-happy child by encouraging the children to dance, be happy, and perform for the tourists.

The children in the school play the role of poor-but-happy child, whilst the tourists play the 'white saviour' role. Both roles constitute main characters; however, the white saviour plays the role of the protagonist. A protagonist experiences an event which allows for character development. Chapter 7 explored the role of the white saviour in Matopo School. The white saviour has multiple forms and shifts as needed. These roles are all different but are underpinned by discourses of Development which position Majority World knowledges as inferior to Minority World knowledges. Further, discourses of Development constitute recipients of aid as passive and grateful. The positioning of recipients as grateful was discussed in the context of funding provided by external companies and the tourists who visit the school.

Given that this tourism does not seem to be disappearing or losing its market hold as of yet, there are some practical ways forward for tourism operators and schools engaging with tourism operators. First, schools should be given the choice of timing and tour arrangements. It is important for schools to have autonomy around the schedules and interruptions of tourism, particularly as student learning should be considered the core business of a school's work (Smithers & Ailwood, 2022). In the case of Matopo School, the visits occurred at a time that was not conducive to children's learning as it was during timetabled lessons. Partnerships that are mutually beneficial should be at the very core of this type of tourism.

Second, tourism operators should think carefully about the images they portray and the way that their tourism continues to encourage outdated and paternalistic images of children and communities. Matopo School was visited by Education Tours, a company that reported to focus on education and learning as part of their core messaging. Despite the guides, such as Greg, hoping to portray an image of Africa that was multidimensional, the tourists who visited the school continued to play into familiar tropes. It seems that although

Education Tours was actively trying to educate and to allay these images, it was not working. The reasons for this are most likely complex; however, one potential reason is that Education Tours continues to rely on images of children in the advertising and on tropes of saviourism in their descriptions of tours. These descriptions and images are a callback to the familiar tropes I have outlined throughout this book.

Third, it is important for tourism operators and for schools to be realistic about the extent to which school tours can influence student learning. Elsewhere, I have discussed the school tour and the repetitive and rudimentary questions that children are asked (Smithers, 2022). In Chapter 6, I detailed the character that children at Matopo School are required to play: the poor-but-happy child. This role does not enable students to learn from tourists, beyond simple questions about their age and siblings. For school visits to be meaningful and productive for learning, tourism companies should consider implementing an educational programme for visitors to undertake.

As this book details, there seems to be no perfect solution to the problem of under-resourcing in schools in Zimbabwe. Global aid and development are predicated on deeply situated notions of the white saviour and images of the poor-but-happy child. In understanding the school tour as part of these broader histories, it is almost impossible for tourism companies and schools to consider an alternative to the current format. The conditions of possibility for a school tour are such that the tour is almost irrevocably tied to the image of the white saviour and to images of needy children. It seems that tourism companies are unwilling to donate money to schools in the areas they visit without some form of reciprocal exchange in the form of school tours. Indeed, this situation itself is tenuous as Thompson explained:

> I hope it is not going to get commercial. I would hate the day when people say, 'well its very commercialised'. And it is going to happen, when the kids speak fluent English and they are not struggling to say anything. People will say, well it looks like the school is well off. That is one of the risks that we actually, yeah, are going to run into. At some point, I think people will say, 'well you got everything, you do not need help'. But that is one of the disadvantages of success. Something will give in some day.

Indeed, when a school appears to be too 'developed' for a tourism company, funding is potentially withdrawn and the school must continue without tourism philanthropy. It is concerning that this is the reality for schools in Zimbabwe, in which they must work within the very specific inclinations of tourism companies to fund their students' education. We may not know what this cycle looks like, but it is sure that there will be negative ramifications for the schools involved – it is an imperative that tourism companies and schools work together to create more sustainable funding arrangements that do not rely on interactions with children.

## References

Aitchson, C. (2001). Theorizing Other discourses of tourism, gender and culture: Can the subaltern speak (in Tourism)? *Tourist Studies*, *1*(2), 133–147.

Ministry of Primary and Secondary Education. (2019). *2018 Primary and secondary education statistics report*. Ministry of Primary and Secondary Education. Retrieved November 10 from http://mopse.co.zw/sites/default/files/public/downloads/2018%20 Annual%20Statistics%20Report.pdf

Smithers, K. (2022). 'What is your name, where do you come from, what is your grade?' Using art-based interviews to highlight the experience of children hosting school tours in Matabeleland North, Zimbabwe. *Tourism Recreation Research*, *48*(6), 912–924. https://doi.org/10.1080/02508281.2022.2133812

Smithers, K., & Ailwood, J. (2022). Developmentourism and school tours in Zimbabwe. In M. Novelli, J. M. Cheer, C. Dolezal, A. Jones, & C. Milano (Eds.), *Handbook of Niche Tourism* (pp. 345–356). Edward Elgar.

# Index

Pages in *italics* refer to figures and pages in **bold** refer to tables.

For Product Safety Concerns and Information please contact our EU
representative  GPSR@taylorandfrancis.com
Taylor & Francis Verlag GmbH, Kaufingerstraße 24, 80331 München, Germany

www.ingramcontent.com/pod-product-compliance
Lightning Source LLC
Chambersburg PA
CBHW060313220326
41598CB00027B/4317